K[...]
WHO
INVENTED
CHAMPAGNE

THE KNIGHT WHO INVENTED CHAMPAGNE

How Sir Kenelm Digby developed robust glass bottles – *verre Anglais* – which enabled wine and cider-makers to produce bottle-fermented sparkling wines and ciders

STEPHEN SKELTON MW

Published by the Author
April 2021

Copyright © Stephen Skelton 2021
22 Radipole Road,
London,
SW6 5DL.

E-mail: mail@stephenskelton.com
Telephone: 07768 583 700

The right of Stephen Skelton to be identified as the author of this work has been asserted by him in accordance with the Copyright, Designs and Patents Act 1988.

All rights reserved. Apart from any use permitted under UK copyright law, this publication may only be reproduced, stored or transmitted, in any form, or by any means, with prior permission in writing of the publishers or, in the case of reprographic production, in accordance with the terms of licences issued by the Copyright Licensing Agency.

A CIP catalogue record for this book is available from the British Library.

1st Edition published in paperback in Great Britain in 2021
ISBN: 978-1-9163296-2-1

Design and typesetting: Prepare to Publish (www.preparetopublish.com)
Index: Dr Laurence Errington (www.errington-index.demon.co.uk)

Every effort has been made to fulfil requirements with regard to reproducing copyright material. The author and publisher will be glad to rectify any omissions at the earliest opportunity

The cover photograph is of Sir Kenelm Digby by Sir Anthony van Dyck (ca 1640) by kind permission of, and payment of a fee to, the National Portrait Gallery, St Martin's Place, London, WC2H 0HE.

Other books by S P. Skelton:
Viticulture – An introduction to commercial grape growing for wine production, 2nd edition
Vine varieties, clones and rootstocks for UK vineyards
UK Vineyards Guide
The Wines of Great Britain (Infinite Ideas Classic Wine Library)
Wine Growing in Great Britain 2nd Edition

Acknowledgements
Thanks go to the following who have provided help and advice during the writing of this book: Rupert Baker (Royal Society), Susan Keevil, Charles Hajdamach, Charles Luddington, the late Sam Mellick, and Tom Stevenson. I would also like to thank the various authors of books, leaflets, pamphlets, PhD dissertations, and multiple websites whose knowledge and information I have trawled in order to write this book. Some are acknowledged in the text. All are acknowledged in the bibliography.

Contents

Preface	7
Principal characters in this tale	9
Timeline	15
Prologue	17
Introduction	21

Chapter 1 — 23
Sparkling wine — 23

Chapter 2 — 27
Mediaeval winemaking — 27

Chapter 3 — 33
The wine trade in Britain — 33
Wine imports — 34
The *négociant* system — 39
Bottling — 41

Chapter 4 — 45
Glassmaking – an introduction — 45
Glassmaking in the British Isles — 51
Window glass — 53
Glassmaking develops — 54

Chapter 5 — 59
Glassmaking – the change from timber to coal — 59
Glassmaking patents — 61
Sir Robert Mansell (1573–1656) — 64
Mansell and glassmaking — 66

Chapter 6 — 71
Glass bottles — 71
Glass bottles in the wine, beer and cider industries — 76

Chapter 7 — 81
Sir Kenelm Digby (1603–1665) — 81
Digby gets married — 86
The voyage to Scandaroon — 88
Digby after the voyage — 89
'Sir Kenelm Digby first invented the glass bottle' — 90
Death of Venetia Digby — 94
Digby imprisoned and then into exile — 97
Digby returns from exile — 99
The Royal Society — 100
Death of Digby — 101
The Closet of the Eminently Learned Sir Kenelme Digbie Kt. Opened — 102
Digby's donation to the Bodleian and Harvard Libraries — 103
James Howell — 103

Chapter 8 — 107
Newnham on Severn — 107
Sir John Winter — 110

Chapter 9 — 113

Dr Christopher Merret	113
The Great Plague and the Fire of London	114
Merret's interest in glass	116
Tom Stevenson and Merret's paper on wine	117
Papers on wine read to the Royal Society	117
Merret's revelations concerning adding sugar to wine to make it 'brisk and sparkling'	123
Merret's other writings	121

Chapter 10 — 123

Introduction to champagne	123
Dom Pérignon	125
Champagne after Dom Pérignon	128
Champagne starts to sparkle	131
Glass bottle development	136
Charles de Saint-Denis, Seigneur de Saint-Évremond	138
Blanquette de Limoux – the world's first sparkling wine?	142

Chapter 11 — 143

Mansell and Digby	143
James Howell and Digby	143
Mansell and Newnham on Severn	144
John Colnett and the Attorney-Attorney General's investigation	144
Sir John Winter and Digby	144
Digby and bottle-making	144

Appendix I: Date changes — 145

Appendix II: Glossary of glassmaking terms — 146

Appendix III: Sparkling wine methods — 150

Appendix IV: Digby's 'Powder of sympathy' — 153

Bibliography — 156

Image acknowledgements — 159

Biography - Stephen Skelton MW — 160

Index — 162

Preface

Since 1974 my working life has been taken up with wine, both its growing and its making, in England. I planted my own vineyard in Kent in 1977 and built, equipped and made wine at my winery there for 22 consecutive vintages. This vineyard and winery were then called Tenterden Vineyards, but today (under different ownership) are much better known as Chapel Down. When I started growing and making English wine, the wines were still wines i.e. not sparkling. However, in the early 1980s, a few intrepid growers and winemakers thought that our (improving) climate might just be good enough to start making sparkling wines using the classic champagne varieties: Chardonnay, Pinot noir and Meunier.[1] At first, things progressed slowly, but, as award-winning and world-class bottle-fermented sparkling wines started to emerge, a trickle turned into a stream and today there is a veritable torrent of high-class sparkling wine being produced, year in and year out, by well over 200 different English and Welsh wine producers.

When you make sparkling wine you soon realise that the bottle is the most important part of the whole process. The process starts, of course, like any winemaking: grapes are harvested, pressed and still wine is made. Then the magic occurs that only the right bottle can achieve, at least in the bottle-fermentation method.[2] The still wine has a measured amount of sugar added, around 24 grammes per litre to give it the full sparkle, plus a yeast culture and is then immediately bottled and closed with a crown cap (such as you might find on most bottles of beer). The sugar and yeast make sure that the secondary fermentation gets under way and your previously still wine starts to sparkle. The actual secondary fermentation takes only a few weeks, but then the ageing process begins. This can take as little as nine months

---

[1] Meunier is the correct name for what many call Pinot Meunier and is officially registered as such on the VIVC (Vitis International Variety Catalogue).
[2] There are at least six different ways of producing sparkling wine of which bottle-fermentation is but one, albeit the one that produces the highest quality wine. An explanation of the different methods can be found in Appendix III.

(although for champagne, 15 months is the minimum and for prestige cuvées and for vintage-dated wines it can be many years) and turns what started out as quite a thin, high-acid wine into something much more glorious and celebratory.

Central to the conversion of still wine into sparkling is the ability of the bottle to hold up to six atmospheres of pressure (around twice the pressure of most car tyres) and keep that pressure from escaping during the whole process without splitting, cracking or breaking. Remember also that in most wineries of any scale, bottles of sparkling wine go through many operations, all of which require the bottle to be man- or machine-handled, sometimes quite roughly. Bottles therefore have to be made to exacting tolerances so that they do not fail during any part of that journey and end up in the consumer's hands in good condition.

However, it wasn't always so. Glass bottles for holding liquids such as beer and wine started to appear at the end of the sixteenth century and it may surprise people to know that Shakespeare, who died in 1616, only mentions the word bottle in relation to wine and beer 13 times in the million words he wrote. Samuel Pepys, on the other hand, who wrote a million and a quarter words in his celebrated diaries between 1659 and 1669, mentions bottles of wine and beer, plus bottles of many other drinks, over 80 times. In between the death of Shakespeare and the Restoration of the monarchy in 1660, glassmaking and, with it, bottle-making, underwent massive changes. These changes, which this book sets out to chronicle, resulted in what many have called 'the invention' of a glass bottle reliable enough to take the pressure of sparkling wines and ciders by Sir Kenelm Digby, the hero of this tale. How true this is, is open to conjecture and this book also sets out the knowns and the unknowns about this 'knight who invented champagne'.[3]

The book also deals with the origins of sparkling wine. Was it, as many seem to think, Dom Pérignon who put the bubbles into champagne? Almost certainly not. For one thing, wines from the Champagne region were not sparkling at the time he was involved with viticulture and winemaking (1668–1715) and the French didn't perfect the making of bottles capable of holding the pressure of a truly sparkling wine until the mid-1770s, almost a century after stout bottles made of *verre Anglais* were available and in use in Britain. I hope you enjoy this book and its cast of characters.

Stephen Skelton MW

Fulham 2021

---

[3] 'The Knight who invented Champagne' is adapted from the title of a popular song called 'The night they invented Champagne' which was in the Oscar-winning 1958 musical film *Gigi*, with words and music by Lerner and Loewe. *Gigi* is probably best known for the very popular, but non-PC, song, 'Thank heaven for little girls' made popular by Maurice Chevalier.

Principal characters in this tale

Martin Luther (1483–1546)
When Luther started the Reformation by nailing his 'protestations' to the church door in 1517, glassmaking became associated with Protestantism, most notably with the Huguenots (the French term for Protestants), whose skills included both glassmaking and silk weaving. Many of these highly-skilled artisans found their way to England, where they could more easily practise their religion and their crafts.

Henry VIII of England (1491–1547)
The annulment of Henry VIII's first marriage to his late brother's widow, Catherine of Aragon, in 1533, triggered England's split from the Church of Rome, and thereby created the Protestant Church of England. This led many European dissenters from Catholicism, including the Huguenots (and others) skilled in glassmaking, to view England as a safe haven.

James I of England (James VI of Scotland) (1566–1625)
James I's 'Proclamation Touching Glasses' of 23 May 1615, which forbade the use of timber for heating furnaces for glass and other processes that needed sustained heat because the timber was wanted for shipbuilding, changed glassmaking overnight. It went from being a transient cottage industry, always having to move furnaces to follow the fuel, to one that could put down roots, build bigger, better, and permanent furnaces, ones on sites that could be supplied with sea- and river-borne coal. The change of fuel and the building of better furnaces improved the quality of glass and made it suitable for windows, both in buildings and in carriages, and for containers for all kinds of everyday products.

Admiral Sir Robert Mansell (1573–1656)
Sir Robert Mansell was one of James I's most successful admirals, who gave up seafaring to become an entrepreneur. It was he who persuaded James to issue the 1615 edict. He owned coal mines, colliers (ships) and glassworks. He held the patent

to make glass using coal, and licenced others to start their own glassworks, paying him fees for the privilege. Mansell and Digby both served on the Navy Board and had numerous other connections.

Sir Kenelm Digby (1603–1665)

Sir Kenelm Digby was a favourite of both James I and his son, Charles I. He knew Mansell well and they served together on the Navy Board. Digby had a lifelong interest in glassmaking and, it is said, had an interest in a glassworks at Newnham on Severn which operated under licence from Mansell. Digby was an 'Original Fellow' of the Royal Society and served on its first council. He was a friend of Christopher Merret's and owned a copy of Merret's translation (with additional notes) of Antonio Neri's *The Art of Glass*. He was appointed to be Queen Henrietta Maria's Chancellor in 1645.

Sir Kenelm Digby by Sir Anthony van Dyck

Dr Christopher Merret (1614–1695)

Christopher Merret was a doctor, born in Winchcombe in Gloucestershire, who became well known amongst learned and scientific circles in London. In 1662 he translated from Latin to English Antonio Neri's *The Art of Glass*[4] (written in 1611) which was the foremost (quite possibly the only) technical book on glassmaking in all its forms and added substantial notes to the original. It is quite obvious that he was familiar with all aspects of glassmaking. He was also an 'Original Fellow' of the Royal Society and read a paper to the members on 17 December 1662 titled *Some Observations Concerning the Ordering of Wines* in which he stated that 'our wine coopers' added sugar to wine to make it 'drink brisk & sparkling'.[5]

---

[4] The full title of Merret's translation is: *The Art of Glass, wherein are shown the wayes to make and colour Glass, Pastes, Enamels, Lakes, and other Curiosities. Written in Italian by Antonio Neri, and translated into English, with some Observations on the Author. Whereunto is added an account of the Glass Drops, made by the Royal Society, meeting at Gresham College.*

[5] The words 'brisk' and 'briskness' were often used at this time instead of the words 'sparkle' and

Principal Characters

Charles I of England (1600–1648/9)[6]

Charles Stuart, the second son of James I, who became Prince of Wales after the untimely death in 1612 at the age of eighteen of his elder brother Henry Frederick, was on the throne from 1625 until his execution on 30 January 1648/9. He was acquainted with most of the principal characters in this tale, as was his wife, Queen Henrietta Maria (see below).

Charles I on horseback. Charing Cross, London

Charles II of England (1630–1684/5)

Charles II, eldest surviving child of Charles I and Queen Henrietta Maria, returned from exile in France and the Netherlands in 1660 at the request of Parliament to reclaim the throne. After the relative austerity of Cromwell's time, the Restoration was a time of furious activity in both the arts and science with the King a great supporter of both. The Royal Society, some of whose members play a part in this tale, received its first Royal Charter in 1662. He died, aged fifty-two, leaving behind no legitimate children, but at least a dozen illegitimate ones, many of whose descendants today are dukes and earls.

James Howell (ca 1594–1666)

James Howell was the steward (manager) at Sir Robert Mansell's Broad Street, London glassworks in 1616. Between 1618 and 1621, he was travelling in Europe, attempting to learn the secrets of Venetian and other glassmaking. In 1622 he travelled to Madrid where he met Sir Kenelm Digby, who helped him recover from an injury sustained whilst attempting to break up a duel. Howell's tutor at Oxford was Mansell's nephew.

'sparkling' in describing wine and ciders. Sometimes (as here) the two words are used together.
[6] See Appendix I for an explanation about changes to dates during this period.

The Knight Who Invented Champagne

Lord Scudamore's Chesterfield Flute

Sir John Scudamore (1601–1671)

Scudamore was a diplomat and cider-maker. His claim to fame (in this respect) is the development of the 'Scudamore Crab' a clone of the Redstreak cider apple. His cider became famous and was known as *vin de Scudamore*. The 'Chesterfield Flute' which is held by the Museum of London, is a magnificent drinking glass engraved with 'the Royal Arms and the arms of the Scudamore family (three stirrups) within lozenge shaped escutcheon linked by festoons of fruit and flowers, with below, a stag beside a gate and five trees (three lopped and one a stump) and the letter S (for Scudamore) repeated three times'. The flute dates from around 1650. Scudamore was ambassador to France 1635–1639 and is said to have entertained Sir Kenelm Digby to dinner.

Rev. John Beale 1608–1683 (or 1603–1683?)
Various sources give different dates of birth

John Beale, a scientific writer and clergyman and said (by himself) to have had a photographic memory, was born in Herefordshire of wealthy farming stock. He was educated at King's School, Worcester and then Eton, after which, in 1629, he went to King's College, Cambridge where he taught philosophy. He was a Fellow of King's College from 1632 to 1640. On 10 December 1662, Henry Oldenburg, the first Secretary to the Royal Society, read a paper to the newly formed Royal Society that Beale had written entitled 'Aphorisms Concerning Cider' which stated that cider-makers often added 'two or three Raisins into every Bottle … and as much as a Wal-nut of Sugar' to improve the cider. On 7 January 1633 he was declared an honorary member of the Royal Society, becoming a Fellow on the 21st of the same month. He was appointed Chaplain to Charles II on 3 March 1665. Both Merret and Digby knew Beale and were quite possibly in the audience at the reading.

Sir Paul Neile (1613–1686)

Sir Paul Neile (also spelt Neil), was the son of the Archbishop of York, friend of Charles I and was knighted in 1633 at the age of twenty. After taking a degree at Pembroke College, Cambridge (he graduated in 1631), he became the Member of Parliament for Ripon in Yorkshire in 1640. He was also the MP for Newark 1673–1677. He was a noted astronomer, owning a telescope used by Sir Christopher Wren who was interested in Saturn and its rings. In 1658 Neile gave a 35-foot (almost 11 m) telescope to Gresham College. He was one of the 12 'Founder Fellows' of the Royal Society and was considered to be the Society's astronomer and telescope builder. On 8 July 1663 he read a paper at the Royal Society titled a 'Discourse on Cider' and in it said that in every bottle you should put a 'little piece of white sugar' which 'will set it into a little fermentation, and give it that briskness which otherwise it would have wanted'.

John Colnett (Jean Colinet)

John Colnett was a glassmaker of Flemish Huguenot origin who had applied for and received letters patent for a period of 14 years from 6 September 1661 for the: 'perfection of making of glasse bottles and of glasse vessel for distillation called bodies heades and receavors of a peece being their owne invention and never before done or used in these our kingdomes'. The granting of this patent was challenged by four glassmakers (see below) who swore before the Attorney General that Colnett had made a false claim, and that it was Sir Kenelm Digby who was the inventor some thirty years before.

Edward Percival, William Sadler, John Vinion and Robert Ward

In 1661/2, the above four, who were all glassmakers once employed by Digby, swore under oath before the Attorney General, refuting the claim of John Colnett that he had perfected the manufacture of glass bottles and said that this had already been done by Sir Kenelm Digby, 'some thirty years since'. On this evidence, the Attorney General, who had made a 'long and serious examination' of the claim, certified that Colnett was not the inventor, and that 'the making of glass bottles is no new invention, for it has been of trade and public use for nearly thirty years'. Colnett's 'special licence' was annulled on 2 April 1662.

Sir John Winter (ca 1600–ca 1673)

The Winter family were substantial landowners in the Forest of Dean and other parts of Gloucestershire and had interests in iron, coal and forest industries. They were Royalists who had the support of Charles I, who borrowed heavily from them. Sir John Winter knew both Mansell and Digby. He was Secretary to Charles I's wife, Queen Henrietta Maria, 1638–1642 and 1660–1669, the year she died.

Queen Henrietta Maria of England (1609–1669)

Henrietta Maria was the youngest daughter of King Henry IV of France and was married to Charles I in 1625. She was devoutly Catholic and, as such, was unable to be crowned, although she was officially known as Queen Mary. In 1644, as the Civil War was going badly for her husband, she fled to France, and did not return until after the Restoration, when her son, Charles II had become king. She knew both Sir John Winter and Sir Kenelm Digby, appointing the former to be her Secretary in 1638 and the latter to be her Chancellor in 1645. She was also the mother of James II, and grandmother of both Queen Mary (of William and Mary fame) and Queen Anne. She gave her name to the US state of Maryland whose royal charter was granted in 1632, and which was first settled in 1634.

Timeline

1450	Johannes Gutenberg develops printing using moveable type
1517	Martin Luther's proclamations start Reformation
1534	Henry VIII splits from the Church of Rome
1547	Henry VIII dies
1558	Elizabeth I ascends to the throne
1572	St Bartholomew Day's Massacre
1597	Gresham College founded
1603	Elizabeth I dies
1603	James I ascends to the throne
1603	Sir Kenelm Digby born
1611	Antoni Neri publishes *The Art of Glass* in Latin
1615	James I bans the use of wood for glassmaking
1615	Sir Robert Mansell joins glassmaking syndicate
1616	Shakespeare dies
1623	Sir Robert Mansell gets glass patent
1623	First appearance of 'string-rim' on bottles
1625	James I dies
1625	Charles I ascends to the throne
1632-3	Sir Kenelm Digby said to have invented glass bottle
1633	Venetia Digby dies
1634	Customs records show 'English bottles' as export goods
1642–51	Civil War in England
1649	Charles I executed
1649–60	Commonwealth of England
1660	Restoration of the monarchy; Charles II regains the throne
1660	Royal Society founded; Royal Charter granted in 1662
1661	John Colnett claimed to have invented the glass bottle; claim dismissed
1662	Beale and Merret present papers to the Royal Society on cider and wine
1665	Sir Kenelm Digby dies

Prologue

The wine arrives from France

It's 1632 and it's cold, very cold. King Charles I is on the throne and it's mid-February. George Blankpain is waiting outside the Rummer Inn in Newnham on Severn for the dray that is bringing eight barrels of wine up from the docks. He knows they will arrive soon as he was down at the docks last night and saw the barrels being unloaded from the Albion, *the ship that brought the wine from La Rochelle. George saw the wine being unloaded onto the dray and saw the ship leave on the flood tide. It will be back in four weeks with another load of wine from France. Soon he hears the rumble of steel-shod cartwheels on cobbles and knows his barrels are just round the corner.*

George is seventeen and has been working at the Rummer for almost two years. He's not a native of Newnham but comes from Glasshouses, a small village some 24 km (15 miles) away. His father, Silas Blankpain, who comes from a Huguenot family that have long been glassmakers, is a friend of the Rummer's landlord, Albert Jones. Mr Jones is a lazy man and was happy to hand over the cellar work to young George and, so far, has no reason to be unhappy with his work. George is always on hand – he would be as he sleeps most nights on the floor of the inn's main bar, making sure that the fire stays alight in the winter and that the place is secure.[1]

The cellar of a busy inn like the Rummer is not a bad place to work. Cool in the summer and warm(ish) in the winter and with no one to oversee him, George is his own master. As long as he makes sure the beer is fresh, checking each barrel as it arrives from Kemp's brewery and making sure that the barrels are marked with the date and used in strict rotation, Mr Jones has no complaints. George also has to look after the ciders, the meads and – his favourites – the wines.

George has decided that wine is strange. Although he has worked at the Rummer for only two years, he knows that wine varies from barrel to barrel and that this year's wines are very different from last year's. He has already had one delivery of the new wines, those from the 1631 vintage, and they have been well received. Jones buys two types of wine: the light red wines known as 'Clairet' from the Gironde and his white wines (in fact they are slightly grey – like the

---

[1] The Blankpains – George and Silas – and Albert Jones are fictional characters.

eye of a partridge) from Sillery, a village in the Champagne region near to Paris.

The beer, of course, comes straight from the barrel – they are lined up behind the bar – and drawn from each barrel through a wooden tap directly into pint or quart tankards made of pewter or into serving jugs which the serving girls take out to the customers. Some of the women don't like drinking out of pewter, so there are some silver-rimmed horn half-pint beakers which he keeps for the daintier clientele.

Cider and wine always used to be served like that, but in recent years things have changed. George has persuaded Mr Jones to put some of the wines into the new glass bottles that they are now making at the glassworks where his father works. This is one of Admiral Mansell's new coal-fired glassworks and one where Mansell's friend Sir Kenelm Digby has been involved with designing the new kilns. The bottles they are now making are darker, thicker and more robust than the bottles they used to use, and are also of a different design to those previously made. They have a thick base, with a pronounced dimple in the bottom and a long, slender neck which ends with a protruding ring which can be used to secure the bung with string to stop it from coming out.[2] Why the bung would want to come out after it has been pushed home is a perplexing question and one that George has been thinking about.

How did the sparkle get there?

When barrels of wine arrive from the ship, the first thing to be done is to get them onto the baulks of timber that sit on the cellar floor, make sure the bung holes are on top, and let them settle for a day or two. Once the wines have settled, it's time to knock the barrels either side of the bung with the 'flogger', spring the bungs out and taste the wines. Most of the wines are really dry and astringent. The red wines are usually fine, a bit sharp and tannic, but so many customers, at least in the winter, drink these 'mulled' – with the addition of herbs, honey and usually served slightly warm – and so they will be fine. The white wines from Sillery, on the other hand, vary quite a bit. George has noticed that some of them have a little touch of sweetness and are more pleasant to the palate and especially popular with some of the women that drink there. These barrels he marks up with a chalk mark and makes sure that when he bottles these, they are stored in a quiet corner of the cellar and kept for special customers.

It's now June 1632 and not far off midsummer's day. The beer and ciders keep on arriving, although cider from last year's apples is now getting scarce and is of questionable quality. Until the apples are ready to be harvested again in late September and October, the Rummer's customers will have to make do with ale – as long as there is barley there is always ale – mead and of course, wine.

---

[2] Now known by glass professionals and collectors as 'shaft and globe' bottles.

Prologue

George has noticed that some of the 'special' bottles of wine he bottled in the spring from the February delivery – the slightly sweeter ones he set aside for the women – were slightly sparkling and that when the twine securing the bungs was cut, the bungs flew out with some force and the wine foamed out of the bottle and sparkled in the glass when they were poured. He had asked his father why this might be, but in his mind he knew. It must have something to do with the touch of sweetness in the wines he had put aside. He hardly dare suggest this to his father, but plucking up courage he did. 'Well, we can soon tell if that's the case,' he said without much conviction, 'go and get some lump sugar from the kitchen and put a small piece in some bottles when you are next bottling and we will see what happens.'

At the next bottling, he did as he was told. Selecting the best bottles, those without any visible cracks or faults and with a good ring around the neck, he popped a piece of lump sugar about the size of a walnut into each of a dozen bottles and waited to see what would occur. As it happens, it didn't take long. With a July heatwave under way, the cellar warmed up a few degrees and within two weeks of the bottling, George walked into the cellar to see three of his 'specials' split in half, four without bungs and five still intact. He thought he ought to open one of the good ones to see what had happened to the wine. Taking out his pocketknife, he started to cut the twine but before he was halfway through, the bung flew out and the wine foamed out of the bottle a yard into the air. He rushed up to find Mr Jones. 'Taste this and see what you think,' he said. Selecting one of his precious Venetian flute glasses off the back of the bar (where, as far as George knew, they were never disturbed), he poured a glass of the sparkling, foaming wine and put it to his lips. 'My God, George, you have got something there – it's delicious. You get back down that cellar and get bottling. We've got a winner – sparkling wine!'

Dutch shaft-and-globe bottle with seal from 1660-1670

Introduction

It is 1615. Shakespeare is still alive and the country is at peace. James I of England (James VI of Scotland) has been on the throne since the childless Elizabeth I died in 1603. He has claimed the throne by virtue of the fact that he is direct in line of descent from Henry VII, his great-great-grandfather. The English Navy, which had been founded as a standing force by Henry VIII and had defended the country from several Spanish Armadas during the Elizabethan era, has been neglected. It needs rebuilding to face both the challenges from the Dutch, French and Spanish fleets, and to secure the trade routes to the new-found lands in the west: the Americas and the West Indies. This means new ships and a requirement for plenty of stout English and Welsh oak. Luckily for James, one of his closest advisors was an admiral, Sir Robert Mansell, who, having given up his naval career and become an industrialist and entrepreneur (as well as a Member of Parliament), saw an opportunity to secure his new-found business of coal mining and glassmaking. Mansell applied to the King to grant him a patent[1] forbidding the use of timber for smelting (mainly iron and glass), and on 23 May 1615 the papers were signed.

The document in question was called 'A Proclamation touching Glasses' and it prohibited the use of all wood and wood products (charcoal) in the glass industry. In a very few years, any process that needed massive amounts of heat, of which glassmaking and iron smelting were the main ones, turned them from being industries which had to be near their fuel source – basically in the middle of a forest where wood was abundant – to ones which could settle in one spot, preferably near a port, on a navigable river, or a canal, where fuel – coal – could be delivered by ship. It is worth remembering that nowhere in Britain is further than around 105 km (65 miles) from a tidal port, and most of the inhabited parts of the country are

---

[1] A patent at this time was not quite the same as it is today. A 'letter patent for a monopoly' was often issued in exchange for a 'royalty' which was paid to the Crown. This 'patent' lasted for a set number of years, but if the monarch fell out with the patent-holder, or the patent-holder fell behind with their payments, then it could be withdrawn and assigned to someone else.

under half that distance from a navigable river or a canal. It was this ability to access readily and cheaply a never-ending, high-quality power source of fuel that turned glassmaking from a cottage industry producing rustic *waldglas*[2] into an industrial one that produced reliable, perfectly formed, robust glass for domestic utensils, packaging and windows, the principal uses for glass in those early days.

Thus, with the stroke of his quill, the King started the industrial revolution that turned the British Isles from an agrarian economy, one based upon wool, water power and wind power, to one where coal and steam brought about unimaginable developments in trade and industry. It was following the signing of the 1615 patent that glassmaking in Britain went from a peripatetic business which chased the fuel from clearing to clearing in the dwindling forests to one where the fuel travelled to the kilns. By virtue of the fact that kilns didn't have to move as the wood ran out, they could be bigger and better, brick-built with chimneys and flues, which made the glass more durable and therefore more useable. It was into this exciting, changing world of glassmaking that Sir Kenelm Digby (probably) developed his *verre Anglais* bottles which enabled the production of (lightly) sparkling bottle-fermented ciders and wines.

The revolution in industrial practices and techniques, which every schoolchild knows as 'the Industrial Revolution', is always said to have started in Britain in the middle of the eighteenth century with developments in the wool industry, followed by the cotton industry. The motive power for this revolution was initially water, with water mills in almost every town and village, followed by steam when the need for pumps to keep the ever-deepening coal mines free of water became necessary. Once coal, in which Britain was luckily abundant, and steam power, in which Britain led the world, became established, the revolution took off, turning a little island with an agricultural economy into one with the world's biggest industrial power.

This story is therefore not only one of the creation of a reliable glass bottle, but also of the start of the revolution in manufacturing and production which is at the heart of what Britain became, and whose legacy can be seen in almost every corner of the globe and where English is spoken in one form or another by a third of the world's 7.5 billion inhabitants.

---

[2] The word means 'forest-glass' or 'wood-glass' in German. It was also known as 'weald-glass'.

Chapter 1

Sparkling wine

Nobody needed to invent sparkling wine because carbon dioxide – the gas that accounts for the fizz – is one of the main by-products of an alcoholic fermentation. Allow yeast to start multiplying in the presence of sugar and three main things are produced: carbon dioxide, a small amount of heat, and alcohol.[1] Winemakers wondering whether a barrel of grape must[2] they are hoping has started to ferment will typically hold their ear to the bung hole. If it has started fermenting, however slowly, they will hear the tell-tale 'pop, pop, pop' of carbon dioxide gas being produced.

The practice of making wine from fruit has been one of man's pursuits for at least eight thousand years and possibly longer. Crushed fruit, left in a suitable container and at a suitable temperature will spontaneously ferment using yeasts either on the fruit itself or in the atmosphere. One can imagine the delight of the first caveman or woman who realised that their container (perhaps a sheep or a goatskin?) of rather battered, bruised and perhaps partially crushed berries had produced a liquid that was slightly

Chapel Down is the largest wine producer in Great Britain

---

[1] Strictly speaking, the alcohol produced by the fermentation of fruit juice should be called ethanol.
[2] 'Must' is the word used to describe grape juice before it turns into wine.

sweet, slightly sparkling and possibly quite alcoholic. What was this mind-altering substance and where did it come from? One can also imagine that this news would travel quickly and it would not take long before our inventive caveperson was selecting the ripest, sweetest and easiest-to-crush fruit, placing it into a suitable container (a hollowed-out rock perhaps?) and impatiently waiting for it to turn into wine. Most fruits contain sugar and most fruits can be turned into an alcoholic beverage. However, despite almost all plants having fruit of one sort or another, only a very few have been consistently used by man to produce fermented alcoholic beverages from their juice alone: grapes, apples and pears are probably the most widespread.

Although winemakers have come a long way in thousands of years, the essentials of winemaking have not: take ripe fruit, crush it, ferment it, drink it. OK, so there is a bit more to winemaking than this, and getting it into a bottle so that it is stable and will keep so that it can be enjoyed at some future date takes care and suitable machinery, but essentially the job is the same. The production of sparkling wine differs from the production of still wine in that during fermentation, whether a primary fermentation or a secondary fermentation (where additional sugar and yeast are added to still wine), the carbon dioxide naturally produced is captured in the container, a tank in the case of Charmat method sparkling wines or the bottle in the case of bottle-fermented sparkling wines.

Whether tank-fermented or bottle-fermented, the one thing that is common to *all* sparkling wines is the requirement to have a bottle strong enough to withstand the pressure created by the carbon dioxide dissolved in the liquid. Whether this is the lightly sparkling (or *frizzante*) *Moscato d'Asti* which typically has only 1 atmosphere

---

Different methods of sparkling wine

For properly sparkling wine, there are in fact at least six different methods of making it:
- Carbonation
- Tank, Charmat or Cuve Close method
- Russian continuous method
- Méthode rural, méthode ancestrale, or péttilant-naturel methods
- Transfer method (transvasage)
- Bottle fermentation, Traditional Method

See Appendix III for more details.

(atm) of pressure, the slightly more sparkling *Asti Spumante* with 3 atm or champagnes and other 'full strength' bottle-fermented sparkling wines which have a pressure of 5–6 atm (around 75–90 lbs per square inch, or psi), the requirement is the same: a bottle that has the strength to withstand the pressure during fermentation and during the often-arduous journey from winery to customer. This can encompass shocks to the bottle during production and transport including rough handling and changes in temperature, all of which might cause a bottle with weaknesses to crack and explode – not what you want for a liquid product packed in glass.

The story of the development of sparkling wine is therefore not so much one of how the wine was made sparkling – this was probably understood by the earliest winemakers – but how a glass bottle was developed which would withstand the pressures involved.

Chapter 2

Mediaeval winemaking

Before science started to creep into winemaking, historically quite recently – probably around the 1850s – the condition of wine was very variable. Until the nineteenth century, if vineyards were treated against pests and diseases at all, it was with plant-based 'teas' which in all probability were of limited effectiveness. Vines were planted very close together, trained to single poles, with grapes growing very near to the ground. In these pre-*Phylloxera* days, vines were propagated by 'layering'[1] and planted in what is known as *en foule*[2] planting. Fixed rows, fixed vine spacings and trellising with posts and wires did not really become commonplace until after the great planting boom of the 1870s, following

Basket wine press from 16th C

Phylloxera. Grape varieties, as we understand them today, were largely unknown, and producers stuck to the varieties local to their region which had proved themselves suitable for the climate of the region and suitable for the style of wine for which the region was known. In addition, it was actually felt that mixing red and white vines in the same plot and having vines from different varieties growing side by side (as they still do in some Douro vineyards) was beneficial. If one sort failed, then maybe

---

[1] What the French call *marcottage* where a winter cane, still attached to the parent vine, is buried in the ground and allowed to root, thus becoming a 'new' vine. In time the original cane can be detached from the parent vine leaving the new vine as a replacement for one that died.
[2] Meaning 'in a crowd'.

another sort would not. One can imagine what the condition of the grapes must have been like and, in all probability, they were picked before they reached full maturity in order to save them from disease and pestilence.[3] Therefore, wine made from these grapes was itself very variable and in many cases of dubious quality.

The basic problems with wine at this time was that it was produced in many instances from unripe and diseased grapes, and winery equipment was primitive, with wood being used for the construction of presses and all fermentation, storage and transport containers, i.e. barrels. Without electricity, stainless steel, plastics, concrete and refrigeration – all the things that in a modern winery we take for granted – making clean, sound wine must have been a real challenge. Fermentation conditions were basic, with only natural yeasts available and no, or at least very primitive, control over temperatures. Cool cellars are all very well for the storage of wine, but not so good for getting a fermentation going, especially if yeasts are weak and levels of nutrients in the grapes are low. Added to this was climate change. Between around 1300 and 1850, the start of European industrialisation, there occurred what climatologists refer to as the Little Ice Age when winter temperatures were significantly lower than they are today. Between 1400 and 1835 the Thames froze over 24 times. The first 'frost fair' was held on the Thames in 1608 and the last in 1814.[4]

How would wine have been made in these pre-modern times? Red and white grapes, growing in the same vineyard, would be picked at the same time and the concept of separate techniques for red and white wines appears to have been unknown. Most of the wine produced was either white, made entirely from white grapes, or a blend of red and white grapes which made a wine which might be described as a tawny rosé, or grey colour – what the French still call *oeil de perdrix*, or partridge eye. The notion of a separate 'red' wine did not emerge until the best châteaux in Bordeaux started to produce *clairet* – from which we get the term Claret – specifically for the British and northern European markets in the mid-1600s at the earliest. The reason for this is that, without any power in the winery, there were no

---

[3] Before the advent of modern anti-*Botrytis* chemicals in the 1960s, many Bordeaux châteaux and Burgundy producers would pick their grapes before they had achieved full ripeness – what today we might call 'physiological maturity' – in order to save their crops. Natural alcohol levels, i.e. at picking and before the addition of sugar to increase alcohol levels, were often around 10 per cent or even lower, whereas today, 13-14 per cent (or even higher) would be the norm.

[4] The Thames as it ran through the centre of London at this time was much wider than it is today and ran more slowly, allowing the water to freeze more easily. The vertical stone embankments, and the sewers that run beneath them, which speeded up the flow of the river, were not constructed until 1865–70 under the direction of Sir Joseph Bazalgette.

efficient de-stemmers to remove stems (or stalks). De-stemming by hand was always possible (but very time-consuming) using a sort of sieve to remove the grapes, and old winemaking books also show the use of a wooden 'trident' to remove some of the grapes from their stalks. However, except for high-quality (and therefore high-value) grapes, one can safely assume that little de-stemming took place.

Grapes would therefore be brought into the winery, crushed as best as could be achieved using hand implements or foot treading, then loaded into the press, complete with stalks, using buckets and scoops made of wood or leather and then pressed. Presses – some of which still survive from the fifteenth century – would have been simple affairs, with primitive wooden screws providing the pressure, or the counterbalance type in which stone slabs are used to apply the pressure. No doubt good and conscientious winemakers would have realised the benefits of what today we call the 'free run' juice, the juice that flows from the crushed grapes before pressure is applied, and this would have been drained off into barrels and kept apart from the later 'press juice' which, as more and more pressure is applied, gets increasingly bitter and tannic as the grape stalks release tannins and phenols. As there were no mechanical pumps, the juice would have been drained by gravity or siphoned off into barrels where it would have been left to ferment. Old winemaking books also show how winemakers moved wine from barrel to barrel by air pressure using leather bellows. In all probability, no additions or treatments would have been made at this stage. Would sugar or another sweetening agent (honey perhaps?) have been added at this stage? Highly unlikely as sugar was rare and expensive and did not become a household item until well after the development of sugar plantations in Brazil and the Caribbean. Although these began in the 1500s, sugar didn't really become cheap enough and therefore widely available until the mid-1800s. The blockade of French ports during the Napoleonic Wars (1803–1815) by the British forced the French to develop the cultivation of sugar beet in France, breaking the monopoly held by cane sugar producers and resulting in lower prices and much greater availability of sugar.[5]

Keeping things clean and sterile was no doubt one of the greatest problems in wineries at this time. The use of sulphur as a disinfectant and preservative in winemaking (as well as in other areas) was well known to the Romans, although its use was confined to the burning of 'brimstone' (another word for elemental sulphur)

---

[5] Napoleon's Minister of the Interior was one Jean-Antoine Chaptal (1756–1832) who unwittingly gave his name to the practice of adding sugar to grape must to increase the alcohol content – what today we sometimes call chaptalisation. Chaptal, who was a distinguished scientist and agronomist, conducted experiments into growing sugar beet on his model farm at the Château de Chanteloup on the Loire.

and was a laborious and time-consuming business. Sulphur 'matches' were made by coating fabric strips in sulphur dissolved in water which were then placed into barrels (with or without wine in them) and set alight. As the sulphur burnt it combined with oxygen to form sulphur dioxide gas. This was known as 'matching' and as many as 70 matches could be burnt in a 100-gallon (455-litre) barrel in order to infuse the wine with enough sulphur dioxide to keep it sound. The wine had to be treated by degrees, filling the barrel to one-third full, burning 20–30 matches, rolling the barrel around for two hours so that the sulphur dioxide was absorbed and repeating the whole process three times. Once the wine had finished fermenting, it was no doubt racked off into clean barrels and sold as soon as possible. It was now down to the wine merchant to prepare the wine for market.

In the Middle Ages (and indeed until well into the late twentieth century), the task of the wine merchant, the 'vintner', was one nearer to alchemy than winemaking. It was the job of all those engaged in the trade in wine to blend wine in order to improve it, to add things to it to make it better, and to rectify faults of all different types. In short, to do whatever was required to make the wine palatable. The winemaker and the merchant also undertook different tasks in the journey of the wine from vineyard to table. The winemaker made the wine to the point where the grapes were turned into wine, and then sold it as speedily as possible. It was the merchant's task to buy the wine in barrel and deal with after that and this is still very much the rôle played by the *négociant* in the sale of many of the best Bordeaux wines today. The producer gets money to pay their bills and can get on with the job of running the vineyard. The merchant can assemble, blend, clarify, and treat the wine so that it can travel, and then fill it into clean barrels for its onward transport to the next link in the chain: the vintner. The vintner's task was much like that of the merchant. Undoubtedly much wine could be sold immediately it had been landed and arrived at the vintner's premises, but poorer quality wines or wines that had suffered spoilage during their journey would need blending and treating to make them saleable.

Once the wine was ready for consumption, it would be delivered out, in barrel, to the customer. Selling wine in bottles would have to wait until a bottle strong enough, and sufficiently uniform in size to be useable as a sales container, became available. This aspect of wine production is dealt with in subsequent chapters.

Chapter 3
The wine trade in Britain

The British Isles, not being a winegrowing region any note (that is, until the end of the twentieth century), was therefore one of the original markets for wines from the mainland of Europe. For over 1,000 years before the Romans first arrived in Britain, the Phoenicians (the inhabitants of the southern part of what is today Lebanon), who it is said came to the south-west of Britain for its tin, undoubtedly had wine on board their ships as they traded with ports near wine-producing regions in southern Spain (they are said to have founded Cádiz) and no doubt loaded up a few extra amphorae for the thirsty Cornish tin miners they traded with. Wine, olive oil and fish sauces, being unobtainable in Britain, were ideal as trade goods for bartering and were all easily transportable in relatively robust amphorae which would pack well in the hold of a ship and act as ballast.

Amphorae showing how they could be stacked in a ship

The invasion and settlement of parts of Britain by the Romans from AD 43 onwards, saw a huge increase in wine arriving in both amphorae and barrels. Excavations of their forts and villas have shown that wines came from all over the Roman Empire, but principally Italy, France and Germany. Whether it was consumed by Romans only or Britons as well is open to question, although over the 400 years that the Romans occupied Britain, it seems highly likely that it became part of the national diet. Whether or not the Romans successfully grew grapes in Britain and made wine from them is open to conjecture and despite one positive sighting of a vineyard site complete with vine pollen in the ground, no grapegrowing or winemaking equipment, such as is found in other regions where the Romans settled and grew grapes, has ever been found. After the end of the Roman occupation, Britain descended into what is known as the 'Dark Ages' and little is known about wine consumption from these times.

Wine imports

Following the invasion of parts of Britain by William the Conqueror and his Norman knights from 1066 onwards, wine production and therefore, one assumes, wine drinking, started to increase. Twenty years later, when the Doomsday Books[1] were compiled in 1086 and 1087, 42 vineyards were listed together with references to vines and wines in another three places. Ten of the vineyards had been quite recently planted and, surprisingly, only 10 of the vineyards were attached to monasteries, the rest being in the hands of King William's nobles and cultivated – one assumes – to provide wine for their dining tables. However, a handful of small vineyards was never going to keep the many thousands of knights, cavalry, archers and foot soldiers who arrived supplied with wine, and there can be no doubt that wine was one of the first commodities to be imported and traded once the Normans settled and took over the administration of much of the British Isles.

In the Middle Ages, the wine trade between Europe and British ports was conducted entirely by sea and wine was transported entirely in barrels. In England, the ports of London, Southampton, Bristol and Hull – one on each corner of the country – were the major ports of entry for wine and it is estimated that wine accounted for a third of all imports by value. In Scotland, Leith, the port for

---

[1] The two surveys, Little Domesday and Great Domesday, were by no means a complete survey of the whole of Britain and they excluded Scotland, Wales, much of northern England, as well as the cities of London and Winchester. In some of these places there may well have been vineyards.

Edinburgh, had been importing wine since at least the twelfth century and what was known as the 'Auld Alliance' between Scotland and France was partly because of, and based upon, the considerable trade in wine between Bordeaux and Leith. It is said that in 1770, Leith's glassworks were producing one million bottles a week from kilns that were up to 30 m (98 feet) high and 12 m (39 feet) in diameter. The trade in wine in barrels for bottling in Leith's numerous cellars continued until the 1970 vintage.

Whilst vineyards on the mainland of Europe were already developed in regions not near to the sea – Burgundy would be a prime example – the ones that were able to trade with distant markets most readily were those on or near rivers along which wine could be transported. Thus, German wines came from the Mosel and the Rhine via Rotterdam and other ports in the Low Countries; and French wines reached Britain via the Loire, the Charente, the port of La Rochelle, and through (though not initially from) Bordeaux with its huge hinterland of vineyards serviced by the rivers Lot, Garonne, Dordogne and Tarn. Wines also came from as far inland as Cahors which is over 300 km (186 miles) from the sea by river. Further afield, wines came from Portugal's Douro region via Oporto and Lisbon, and from Jerez and Malaga in Spain. From even further afield, wines came from the Canaries and Madeira.

We know that the supply of wine greatly increased following the marriage of Henry, Prince of Wales, to Eleanor of Aquitaine in 1152, who brought with her the Angevine Empire which included the rich wine-producing regions of Bordeaux and its hinterland, plus Gascony and much of the Loire. This led to an influx of Gascon wine merchants to London, who came to dominate the wine trade, although it did become a two-way trade. Libourne, situated at the confluence of the Dordogne and Isle rivers, was founded by an Englishman, Roger de Leybourne (from the village of Leybourne near Maidstone in Kent), in 1270 in order to gain control of the wine that passed through it and collect taxes due. In 1338, Edward III taxed wine for the first time, needing to repay the sums he had borrowed to give to his foreign allies to fight for him in the war we now call the Hundred Years War.

Over the next 100 years, the wine trade between ports on the French Atlantic coast – principally Bordeaux and La Rochelle – and Britain increased hugely. The five Cinque Ports on the south coast – initially Sandwich, Dover, Hythe, New Romney and Hastings[2] – were given their first 'general charter' in 1260, although

---

[2] When the channel serving New Romney silted up and the harbour could not be used, Rye took its place and Winchelsea was added to the list. Additional 'limbs' were added to the original towns which included Lydd, Folkestone, Faversham, Margate, Deal, Ramsgate, Brightlingsea (in Essex) and Tenterden. A further collection of towns was added to the confederation at a later date.

earlier charters had been given to some of the ports individually in 1155–1156. In exchange for providing the king with 'fifty-seven ships for fifteen days service a year' – in effect creating the first English Navy – the Cinque Ports gained substantial freedom from taxes and an ability to police the coastline and punish offenders (see box). This 'smuggler's charter' permitted seafarers to do pretty much what they wanted without fear of punishment.

The Cinque Ports Charter of 1260

The Charter gave the Cinque Ports: "Exemption from tax and tallage, right of soc and sac, tol and team, blodwit, fledwit, pillory and tumbril, infangentheof and outfangentheof, mundbryce waifs and strays, flotsam and jetsam and ligan". Quite a list!

- Tallage – a form of arbitrary taxation levied by kings on the towns and lands of the Crown, abolished in the 14th century.
- Soc and sac – the right of a lord to hear and decide legal cases on his estate without recourse to other courts.
- Tol and team – privileges granted by the Crown to landowners under Anglo-Saxon and Anglo-Norman law.
- Blodwit – The right to punish shedders of blood.
- Fledwit – the right to punish those who were seized in an attempt to escape from justice.
- Pillory – to be put in the stocks.
- Tumbril – to be put in a ducking stool as a punishment.
- Infangentheof and outfangentheof – privileges granted to feudal lords which allowed them or their servants to execute summary justice on thieves within the borders of their own manors or fiefs.
- Mundbryce – the breaking into or violation of a man's mund or property in order to erect banks or dikes as a defence against the sea.
- Waifs and strays – people or animals, both farm and domestic, found on one's property.
- Flotsam – floating cargo that has come from a wreck.
- Jetsam – floating cargo intentionally discharged from a ship in order to try and save it.
- Ligan – cargo from a sinking ship that has sunk to the sea floor.

CHAPTER 3

In 1214 King John gave the towns of Bordeaux, Bayonne and Dax an exemption from *La Grande Coutume,* a tax on wine which had been in place since 1154. A year later La Rochelle and Poitou were added to this list. The tax concession was given for the support of these towns against Phillip II of France. In 1224 Bordeaux received privileged access to the English market through London, and its exports soon dwarfed the production from other French winemaking regions. Merchants bringing wine from the winemaking region which stretched from Bayonne in the south to La Rochelle in the north were allowed to sell their wine in London before St Martin's Day (11 November) and wine from the Languedoc, which could travel up the Gironde (which was navigable for flat-bottomed craft as far away as Toulouse) was forbidden to be exported until after 1 December.

On 15 July 1363 the 'Mistery of Vintners' obtained a Royal Charter from Edward III which gave them the sole right to trade in wine from Gascony, together with other important wine-related privileges. Also in 1363, Sir Henry Picard, Master of the Vintners' Company, former Lord Mayor of London, and King's Butler, entertained the kings of Cyprus, England, Scotland, France and Denmark, thus giving rise to the name 'Five Kings House', the alternative name for Vintners' Hall. It is estimated that at this time England was taking one-third of all wines that were produced in the wider Bordeaux region. The Vintners' Company, which took over from the Mistery, was formally incorporated on 23 August 1437 and oversaw the trade in wine, which, by the middle of the 1400s, made up nearly one-third of all imports into England. The Vintners kept their dominance until 1553, when Edward VI restricted their rights to be the sole sellers of wine.

Until the invention of bottles that were of a standard size, strong enough to stand the rigours of bottling, transport, handling and sale, and cheap enough to be included in the price and widely available, all wine (and indeed almost all other liquids) was transported and sold in barrels. Wine for export would leave the producer's winery in barrel,[3] travel to the merchant's cellars, which would usually be situated at the quayside. Here the wine

> Away, you cut-purse rascal! You filthy bung, away! By this wine, I'll thrust my knife in your mouldy chaps, an' you play the saucy cuttle with me. Away, you bottle-ale rascal! You basket-hilt stale juggler, you!
> - Doll Tearsheet, *Henry IV Part II*

---

[3] The most common barrel size in the Middle Ages was the 'tun' which was around 955 litres (210 imperial gallons) or the 'butt', a barrel of half the size of a tun i.e. 477 litres (105 imperial gallons). Barrel sizes have changed over the centuries and were not standardised until the Imperial System was introduced by the British Weights and Measures Act of 1824.

would probably be placed into larger barrels for blending and further treatments, some of which we would today certainly term as adulteration, but which were then fairly standard practice. Wine was often sweetened with sugar or honey, flavoured with herbs and spices and 'improved' with sweet wine made from dried grapes. Such practices were considered as positives, given that they were to improve wines that had been tainted by dirty barrels, oxidised through lack of preservatives, or which had started to turn acetic. Old winemaking books and cellar-management books all abound with different recipes of making this wine and that wine using all kinds of different ingredients. In Henry IV Part 1, Shakespeare calls Falstaff 'Sir John Sack-and-Sugar' after his habit of sweetening the dry fortified wine known as 'sack' which came from various parts of Spain including Jerez, Malaga and the Canary Isles.[4]

Wine was generally sold with a generic regional name, or a description, rather than the name of an *appellation* or individual property and of course, sold in barrel rather than bottle. Bottling would have to wait until the 1660s. The first mention by name of a wine in England is in the cellar book of Charles II where 'Haut-Brion' is mentioned. In France, Haut-Brion is mentioned by name as early as 21 January 1521, in a document detailing the sale of an annuity which states that the seller is to get 'four pipes of wine from the vineyard of Aubrion' as part of the annual payment. Samuel Pepys,[5] the great diarist of the 1660s, also mentions this wine and says that on Friday 10 April 1663 he went to the

Samuel Pepys. Stained glass window in Woolwich Town Hall

---

[4] Whether 'sack' is a corruption of the French for dry – 'sec' – or a corruption of the Spanish 'sacar', meaning 'to draw wine from a solera', is uncertain. It is generally taken to mean a fortified wine from Spain and its islands.

[5] Samuel Pepys, 1633–1703, was an important figure of this period. He was a Navy administrator, rising to become Chief Secretary to the Admiralty under both Charles II and James II, a Member of Parliament for Harwich, and President of the Royal Society 1684–1686. He is best known, of course, for the diary he kept which ran from 1 January 1659/1660 until 1669 when, thinking he was going blind, he stopped keeping a daily diary. The diary covers the period during which London suffered both the Plague and the Great Fire and contains much about how the good and the great behaved, as well as ordinary people going about their daily lives. He is especially good on eating and drinking.

'Royall Oak Tavern in Lumbard Street' and 'here drank a sort of French wine called Ho Bryan[6] that had a good and most particular taste that I never met with'. As Thomas Jefferson found as he travelled around Italy and France in 1787 during the period that he was the American ambassador to France (1784–1789) it was always possible to order wine from a producer whose wines you admired. Amongst other wines he collected on his long journey, he ordered 250 bottles of 'sweet white Frontignan' from Monsieur Lambert to be shipped to Paris. Jefferson was, of course, most famous (that is in relation to wine – he was, of course, the third president of the United States) for owning several vintages of Bordeaux First Growths, some of which met a less than glamourous end. A bottle of 1787 Château Lafite, with the name of the wine and Jefferson's initials etched onto the bottle, was bought by billionaire Malcolm Forbes for about $157,000 in 1985, then a world record for a bottle of wine. It was displayed in a glass case under lights, but became worthless when the dried-out cork fell into the wine and nobody noticed in time to at least taste it before it was ruined. Other bottles of supposedly Jefferson First Growths were sold to another American billionaire, Bill Koch, and these turned out to be fakes, with the dates on the bottles recently etched with an electric dentist's drill.

The *négociant* system

The *négociant* system in Bordeaux, through which much wine of all qualities still flows, started in 1620 when the Dutch firm of Beyerman was founded. This was, of course, well before bottling as a means of transporting and selling wine was developed. However, as techniques of bottle making improved, and bottles of sufficient quality and uniformity (of both size and shape) were available, the practice of both bottling at source and bottling at the *négociant's* premises became more widespread. Charles Ludington, writing in *Inventing Grand Cru Claret: Irish Wine Merchants in Eighteenth-Century Bordeaux* says that whilst there were English, Dutch, German and Danish *négociants*, by a large measure the majority were Irish, both Catholic and Protestant. The Irish bought all the best wines and shipped them to the northern European markets. For the top-quality wines, London was the main market, with lower quality wine, but in

---

[6] Today's Bordeaux First Growth Château Haut-Brion. It was owned at this time by the Pontac family whose son, François-Auguste, was sent to London in 1666 to open a restaurant, grocers and tavern which was named the 'Sign of the Pontac's Head'. It became a very fashionable place to meet, eat and drink and survived until 1780. Jonathan Swift is said to have complained that the wine was 'dear at seven shillings a flagon'.

much larger volumes, going to Dublin and Leith. Merchants such as the Bartons, the Johnstons and the Lawtons were all of Irish origin and all set up shop in the mid-1700s. Ludington says that these merchants developed both the taste of and the market for 'claret' and by regularly blending thin Bordeaux wines with stronger red wines from the northern Rhône (Hermitage) and wines from Alicante and Benicarló (a town midway between Barcelona and Valencia) they created a style of wine that wine drinkers liked.

Ludington has detailed the blending practices which commonly took place in order to satisfy the demands of their northern European customers. He discovered an old warehouse ledger which showed that wines such as the Lafite 1837 contained wines from seven other châteaux, plus a dollop of 1840 Hermitage to round it out. In some instances, it was possible to order a blend of wines from two or more top châteaux, if that's the style you liked. The 'hermitaging' of wines was commonplace and led in some cases to Medoc vineyards actually including up to 15 per cent Syrah vines in the *encépagement* of their vineyards. Wines from other regions and even other countries found their way into 'clarets' blended by *négociants* for their customers, with robust and alcoholically strong Spanish and North African wines proving ideal. Joseph Addison, writing in 1709 in the short-lived *Tatler* magazine, said:

> There is in this city [London] a certain fraternity of chymical [sic] operators who work underground in holes, caverns, and dark retirements, to conceal their mysteries from the eyes and observation of mankind. These subterraneous philosophers are daily employed in the transmigration of liquors, and by the power of magical drugs and incantations, raise under the streets of London the choicest products of the hills and valleys of France. They can squeeze Bordeaux out of a sloe, and draw Champagne from an apple.

The job of the *négoce* was severalfold: to buy wine from the châteaux soon after the harvest in order to help the producer's cashflow; to buy every wine from the châteaux, good and not so good, as they had the facilities to treat, blend and – frankly – adulterate so that they could find a market for them; and to make money. In these three objectives, the *négociants* were very successful. The producers got the cash to settle their bills; the public got wines that were better than they might have been in their raw state; and the *négociants* made money and became rich, some very rich. The *négociants* were aided in their activities by brokers known as *courtiers* who act as middlemen (and yes, they were mainly men) who had their pet producers and, for 2 per cent of the proceeds, arranged the sale and transportation of the wine to the merchant's cellars on the 'Quai de Chartrons'. This is still the practice today.

The novelty of châteaux bottling probably started in the late 1700s with a Châteaux Lafite bottling of 1797, although other châteaux disagree. Maybe it was the 1797 vintage bottled at a later date? What seems fairly obvious is that some of the top châteaux bottled some of their output, typically in years when the vintage was deemed 'good' and therefore in demand, but continued to sell much of their wine, often all of their wine, through the Bordeaux *place*, the system of *négociants* and *courtiers*. The first time a château bottled some of its *grand vin* was in 1869. Châteaux Lafite, which had been bought by Baron James de Rothschild the year before, bottled the 1868 wine which came with the purchase of the property. Following this, '*mis en bouteille au château*' became a sign of quality and provenance which was aimed at giving the wine-drinker confidence in their wine. However, the top châteaux didn't always bottle every vintage, and as the quality, prestige and demand for the wines of Bordeaux increased, the Bordeaux *négociants* were as important to the taste and style of the wines they sold as were the original winemakers.

Bottling

The first producer to bottle all its first-grade wine, its *grand vin*, at the châteaux was the twenty-two-year-old Baron Philippe de Rothschild at Mouton-Rothschild, then only a *Deuxièmes Cru*. He had inherited the estate from his uncle two years previously, in 1922, and decided to bottle the whole of the 1924 vintage at the châteaux, something then unheard of. This helped the wine gain both prestige and publicity. He also (partially) solved the problem of selling less-promising vintages by inventing a second wine, Mouton Cadet, to blend with other Bordeaux wines. This started in 1930. In 1973, Baron Philippe's long-held ambition to raise the estate to join the ranks of Bordeaux's *Premier Crus* was achieved.[7]

Wine continued to arrive in Britain in barrel for bottling and sale long after châteaux bottling was commonplace. London, as well as other wine centres such as Bristol and Leith, were important to the wine trade and in cellars and under railway arches, as well as in purpose-built warehouses, wine was imported in bulk and bottled on demand. The wonderful Victorian brick arches off Tooley Street and below

---

[7] It is widely rumoured that the only reason Mouton was held back in the 2nd category when the 1855 Médoc Classification was drawn up, which was solely based upon the prices achieved for the wines, was that the estate was then owned by the London-born Nathaniel de Rothschild who had purchased it, then called Ch. Brane-Mouton, at auction in 1853.

The Knight Who Invented Champagne

London Bridge Station (recently renovated and re-purposed as part of the new station complex) were once a warren of wine cellars where blending and bottling took place. I recall visiting the premises of one very well-respected City wine merchant in around 1977 in order to look at a hand-corker they wished to sell, and seeing row upon row of dusty, dingy wooden *foudres* (barrels) with name-tags hanging on them – Beaune, Pommard, Volnay etc. – and a few more without tags at all. I have often wondered what sort of blending went on to suit their customers' palates. It is said that one well-known Bristol wine merchant wrote to his Bordeaux suppliers in 1945 asking when they could start re-supplying him with their 'Fine Claret' which was then (as it is today) a mainstay of their red wine sales. 'Not until we can get wine out of Algeria,' came the reply, 'without it, the blend will not be complete.'!

Long-forgotten companies such as the Teltscher Brothers with their Yugoslavian wine *Lutomer Laski Rizling*, Deinhard's *Green Label Mosel* and Sichel's *Blue Nun,* all had substantial London-based bottling empires. When I was setting up my own first winery in 1978, I well remember going to Paddock Wood in Kent to see the new Blue Nun bottling plant, then being constructed. Wine would be imported in rail tankers from their winery in Mainz for bottling in Britain. To start with, this bottling facility was very successful, but within a few years the demand for German and German-style medium-dry white wines started to collapse in the face of competition from Chardonnays and other white varietals from the New World.

The much-loved and almost uniquely British institution, the Wine Society,[8] has a long and interesting history when it comes to the bottling of wine. In 1902 it signed a lease on cellars in Hills Place, just off Oxford Street, which had originally been built in the 1860s by wine merchant George Haig. At 7.5 m (25 feet) deep and five aisles wide, the cellars had been constructed to form the foundations of the building above. Here wines arrived in barrel, were bottled and, after bottling, were stored until required. The peace and quiet of the cellars were somewhat disturbed in 1909 when work started directly overhead on building the London Palladium (which opened with a variety show on Boxing Day 1910). The Society expanded by leasing premises off Joiner Street which then ran under London Bridge Station and led to the warren of cellars already mentioned. When the Society outgrew these premises, additional space was taken in

---

[8] Officially called the International Exhibition Co-Operative Wine Society, it was founded in 1874, following the last of the series of Great Exhibitions when a parcel of Portuguese wine, originally intended for visitors to the exhibition, was discovered in barrels in the bowels of the Albert Hall. Not wishing to offend the Portuguese, a series of lunches was held to publicise the wines, and such was the reception, that a co-operative society was set up to import and bottle wines from around the world to 'sell them at cheap rates to Members of the Society'. Each member was allowed one share, and no dividend would ever be paid. This is still the situation today.

Rotherhithe at the St James Bond (which apparently used to flood at very high tides) until eventually, in 1965, the whole operation – barrel storage, bottling and bottle storage (plus offices and tasting rooms) – was relocated to a greenfield site in Stevenage where it remains to this day. Over time, the freehold of the Stevenage site has been purchased and today occupies almost eight acres and has some of the most modern above-ground wine storage premises in the country. Bottling of some of the Society's wines continued until 1992 when the last bottle of the Society's Crusted Port (said to be still drinking well) was bottled.

Bottling underway at the Wine Society's Hills Place cellars which were under the Palladium. Note the sloping ceiling which is the stalls seating above

Bottling of entry-level and large volume wines has always taken place in Britain and still does. In 1964, the Co-operative Wholesale Society (CSS) opened the first modern bulk-wine bottling plant at Irlam, situated on the Manchester Ship Canal, which is today owned by Kingsland Drinks. At one stage, the premises were designated as part of Spain for the purposes of bottling Sherry which, under Spanish *appellation* rules, may only be bottled in Jerez. Accolade Park (formerly Constellation Park) at Avonmouth, near Bristol, is probably the biggest bottling plant in Britain, bottling the equivalent of 300 million bottles a year, down six lines for both bottles and bag-in-box. Their former managing director told me that the lines ran 24 hours a day, every day of the year, and wine couldn't be bottled more cheaply.

Chapter 4
Glassmaking - an introduction

Whilst we think of glass as a man-made product, it does in fact occur naturally. A type of glass called obsidian is found on volcanic sites, dating back to at least 700,000 BC, and was used by our ancient forbears for making tools. These included knives and arrowheads on account of the way in which obsidian can be shaped by chipping to create a very sharp cutting edge, in a very similar way to how flints were used. There are also other naturally occurring glasses: a type of glass called fulgurite is sometimes produced when lightning strikes a beach; and a glass known as moldavite was discovered at a meteor impact site in Bohemia. And after the Trinity nuclear bomb was tested in the desert in New Mexico in July 1945, a less naturally occurring glass-like substance known as trinite (also called atomsite and Alamogordo glass) was also found. Principally, whenever sand, a suitable flux (known as a 'network modifier', and necessary to make the molten sand flow) and a heat source are present, glass will occur.

Glass has been made by humans for almost as long as fire has existed. Light a big enough bonfire on top of sand, on a beach for instance, and with luck (and especially if some dried seaweed, which acts as a flux, is also on the fire) some glass will be produced. Glass is actually liquid silica (of which sand is one of the main sources) which, when cooled, turns into a solid which is either clear or opaque, and occurs in colours as multiple as those on any colour chart. Because glass has been heated to a very high temperature in order to melt the materials it is made from (1,100°C would be typical, but certain types of glass require higher temperatures[1]), it is unaffected by most acids and other corrosive liquids and is therefore ideal for the storage of many different products, both domestic and industrial. Glass can also be completely clear, making it useful for things that need to be looked through —

---
[1] Borosilicate glass (such as Pyrex and other hard glasses) melts at around 1,650°C)

Isaac Judeus inspecting a patient's urine in a urinal

Shabti of Queen Henettawy covered in faience glass

windows, spectacles, and lenses for instance – and for containers where the contents need to be visible. One of the earliest widespread uses of glass was for small horizontal flasks known as 'urinals' which a physician would use for inspecting a patient's urine.

Glassmaking stretches back many millennia. The Ancient Egyptians (ca 5,500 BC) are generally credited with having started making glass, and coloured glass beads have been discovered from around this time. Whether these were made on purpose or were the by-product of another heat-using process such as copper and iron smelting (where sand might have been used as a furnace base) is not known. Glass-type glazes, known as 'Egyptian faience', were also produced at this time and are often blue or blue-green in colour, possibly the result of being left after the production of making copper for which malachite is the principal source of ore. Glass was a rarity and it is found in jewellery and objects of worship, and thus its makers achieved an important status, above that of other artisans making objects from seemingly base materials.

Once it was realised that glass could be made into objects that would hold liquids, glassmaking expanded and early vases and primitive bottles dating back to 2,000 BC can be found in many places of the world. Glass blowing appears to have been discovered at the end of the BC era and 'hollow-ware' greatly expanded the range of products being made. Ornaments made wholly or partly of coloured glass were also popular and helped glassmaking become something over and above a 'trade', and imbued glassmakers with a nobility that persisted for many centuries.

Ancient glass was made using soda or potash for a flux. Soda comes from natural deposits rich in sodium carbonate or from halophytes (salt-loving plants) such as common glasswort (*Salicornia europaea* also known as samphire), as well as other saltworts. Kelp, an abundantly occurring seaweed, is also rich in sodium carbonate. These plants (and I am including a seaweed as a plant in this context) all absorb the salt in the seawater they grow in and, when burnt, leave ashes rich in soda which were used for both glass and soap making. These plants are also often called *barilla*, the Spanish for saltworts. Potash, often known as kali, short for alkaline or from the Latin word *kalium*, was obtained by burning bracken, a plant found in great abundance in many wooded regions.[2] Lime (calcium hydroxide) is also an important ingredient in glass and many plant ashes also contain lime.

During the era of the Roman Empire, glassmaking was established in northern Italy in and around Venice, especially on the island of Murano which was noted for its crystal-clear *cristallo* glassware. The first spectacles are said to have been invented on the island in the 1290s. Altare, near Genoa, and Bavaria in southern Germany, were also centres

Mould-blown Roman bottle

of glass production. Transport, for both raw materials in – especially fuel, which is required in large quantities – and for finished goods out, was an important consideration for the siting of a glassworks and therefore many of the centres of production at this time were by navigable rivers or near the coast.

As the Roman Empire started to disintegrate and other centres of population started to rise, glassmaking became more widespread (and therefore more fragmented) and some of the old skills appear to have been lost. But, in time, glassmaking centres were re-established throughout Europe, with Alsace and

---

[2] In agriculture, potash is represented by the letter K.

Lorraine in France; the Mosel in Germany; Bohemia in the Czech Republic; and Silesia now mainly in today's Poland, being the most important regions. After the European Reformation, which was started by Martin Luther's publication of his 'protestations' in 1517, glassmaking became associated with Protestantism, most notably with the Huguenots (the French term for Protestants), whose skills included both glassmaking and silk weaving. Following the death of Henry II of France in 1559 and the accession of his son, Francis II, whose wife was the fervently Catholic Mary, Queen of Scots, Huguenot persecution increased and many fled to the Low Countries – today's Holland, Belgium and Luxembourg – as well as to England. The largest exodus of Huguenots occurred following what is known as the St Bartholomew Day's Massacre which took place in Paris and other parts of France between 23 August and 3 October 1572.

Glass, although it is made from a few very basic materials – sand, lime and a flux being the main ones – requires huge amounts of fuel, especially if that fuel is low-energy timber. It has been estimated that each kilogramme of glass required around 150–200 kg (330–440 pounds) of wood to be burnt for fuel, plus, if wood-ash is the flux, a further 65 kg (143 pounds) per kilogramme of glass of suitable wood to provide the ashes.[3] The sand, lime and the flux, if not obtainable locally, could be brought in and stockpiled for use at a later date. Fuel, however, was too bulky to move far, given the primitive nature of both the methods of transport and the roads. Therefore, the furnaces tended to follow the fuel being used and, once that was gone, they were dismantled and moved to an area where fuel was available, i.e. another part of the forest. Glass, although possessing many good, unique properties, is both heavy and fragile, making its transport to the customer both expensive and difficult. It is for these reasons that its introduction and more widespread use was slow. And, of course, glass was not the only product that needed lots of fuel. Iron, lead and other products that needed considerable heat were all in competition for bits of untouched forest.

The traditional furnace for producing *waldglas* was relatively simple. Maybe 5 m (16 feet) in height overall, with a central section holding the crucible at the bottom and a separate furnace for the annealing process. Annealing is where the glass product being produced is gradually cooled until it can be handled. This final process, which in a more sophisticated glassworks takes place in a separate furnace called a 'lehr' (from the German *lehren* meaning 'to teach'), makes sure that the glass

---

[3] There are various references to how much timber was needed in glass and ironworks. One says that 20 imperial tons of iron needed 4,900 cords of wood. A cord is a measure of stacked timber, 4 feet x 4 feet x 8 feet (1.2 m x 1.2 m x 2.4 m), around 128 cubic feet (3.63 cubic metres).

is strong and less likely to 'crizzle' (craze or crack) after cooling. Temperatures in a wood-fired lehr were typically 450–550°C. The crucibles in these early furnaces varied in size. Some excavations show quite small ones, maybe 30–40 cm (12–16 inches) in diameter at the rim, 25–30 cm (about 10-12 inches) high and holding only 10 litres (2 imperial gallons) of molten glass. There might have been several of these small 'pots' in one furnace. In other glasshouses, larger pots have been found: up to 45 cm (18 inches) in diameter and 60 cm (24 inches) high, each capable of holding between 50 and 75 kg (between 110 and 165 pounds) of molten glass. More advanced furnaces were on two levels, with the crucibles for the molten glass to be worked on the lower level, nearest the fire, with crucibles for pre-heating the raw materials above. In these furnaces the heat source, wood (and not charcoal), was positioned at either side of the crucible in a fire-trench, the heat finding its way upwards and around the crucible. The buildings were solid enough and waterproof enough for their purpose, but glassmakers knew that within a few years they would be moving on, so foundations were insubstantial, one of the reasons why only a few of these early *waldglas* furnaces have been found and subjected to archaeological examination.

A steel marver in use during glassblowing

The production of glass is a multi-part process. Firstly, there is the pre-heating of the raw materials to around 900°C, which might take 24 hours, and this is followed by the 'melting' where the temperature is raised to around 1,100°C so that the materials become liquid and fuse together. The molten glass is now called 'metal', which is allowed to cool a little to around 1,050°C until it reaches a semi-plastic state.[4] The glassmaker then collects a 'gob' or 'gather' of molten glass on the end of his (or her – but most usually his) blowpipe (or blowing iron), and then, through a combination of twisting and turning, combined (usually) with blowing, shaping and cutting with specialised tools, turns the raw glass into a manageable shape, then called a 'paraison' (and today a parison). The object being made might well undergo several heatings and coolings during the making process, as well as being rolled on a slab of stone or metal, known as a 'marver', before it is finished. In many instances, once

---

[4] These are typical temperatures, but there are many different types of glass and different types of furnace so temperatures can vary depending on many factors.

The Knight Who Invented Champagne

Shaft-and-globe bottle with deep punt

the object being made has been blown to the desired size, it is transferred from the blowpipe to a solid iron rod called a 'pontil' (or 'punty'). Using a small blob of glass, the item is glued to the pontil whilst any decorations or, in the case of bottles, the string-rim around the top, are applied.[5] Once the item is finished it is broken off the pontil. When the pontil is detached, it often leaves a sharp mark which, in the case of a bottle, can be hidden by pushing the base of the bottle upwards, forming a depression which we call the 'punt'. This is also known by glassmakers as the 'kick-up'.

It should also be noted that glass 'cullet'[6] whether from the glassworks itself in the course of production (known as 'knock-offs' and 'moils') or whether from glass fragments collected from elsewhere and recycled, formed an important part of the materials used in new glass production. Typically, up to 20 per cent of a batch of new glass would consist of cullet as this made the fritting and melting parts of the process easier and saved fuel. It is also the reason why, when former glassworks are investigated, far fewer fragments of what was being made are unearthed than, say, at sites where pottery was being produced (broken pottery makes great hardcore). In addition, finding fragments of, say, glass bottles at the excavation of a former glassworks does not prove that that glassworks made bottles, as the fragments may have been brought to the site as cullet.

Whilst dry timber was the original fuel – woods and forests covering much of the land in many parts of Europe (including Britain) – where there was easy coal and other fuels (such as lignite and oil shale) to be had, glassmakers started to experiment with these. German glassmakers started to use lignite (brown coal) around 1580 but, whilst it burns well (if dirtily), it can never reach the temperature of dry wood. In

---

[5] The word 'pontil' comes from the French for bridge, which is *pont*. Blown glass vessels always start out with the glassmaker using a blowpipe, and finished by transferring it to a solid rod known as a 'punty' rod or iron.
[6] Cullet is fragments of broken glass. The word probably comes from *cueilltte*, French for the act of gathering, picking or collecting.

Neri's book *L'Arte Vetraria* (The Art of Glass), written in 1611 (and translated by Dr Christopher Merret in 1662, of which more later), both coal and sea-coal[7] are mentioned as fuels numerous times, both when used in conjunction with timber or charcoal. This is despite Merret stating on the first page 'To avoid Authors Repetitions, Observe – All the fires must be made with dry and hard Wood'. The French are also said to have experimented with coal in a Rouen glassworks in 1616. It was, however, the banning of wood as a fuel for heating furnaces and smelting iron and other metals in 1615 by James I that kick-started the change from wood to coal and, in doing so, glassmaking was changed forever.

Glassmaking in the British Isles

The lack of archaeological evidence would suggest that glassmaking did not take place in Britain before the arrival of the Romans. However, during the Roman occupation, which lasted for almost four centuries, glass, both decorative and utilitarian, was made and glass artefacts and small window panes cut from spun discs of glass have been found dating from this period. Following the departure of the Romans in around AD 410, glassmaking continued, although how widespread this was is not known. The arrival of Christianity in Britain, which dates from the mission of St Augustine in AD 595, prompted the building of places of worship in which stained glass windows, made of small pieces of coloured glass, were often a feature. At St Paul's Monastery in Jarrow, which dates from AD 686, fragments of coloured glass have been found, and the Venerable Bede, who lived at the monastery, writing in 731, stated that the abbot, Benedict Biscop, brought in glaziers from Normandy. The stained glass which has been found at the site, and dated to the time of the building of the monastery, shows that the glass was actually made from both new and recycled glass fragments which were originally made in the Levant, present-day Syria and Lebanon. However, the glaziers established a glassworks by the River Wear – today the site of the National Glass Centre – and, it is assumed, taught native Britons the art of glassmaking, as glass made from local materials can be dated to this period.

---

[7] Sea-coal has more than one meaning. There are many places in Britain where coal can be found on the beach, either where coal in seams in the cliffs washes out through the action of waves or where sub-sea seams of coal are eroded and washed up onto the beach. However, once coal started to be mined commercially and carried to distant ports by sea, then this coal was also known as 'sea-coal'. Seacoal Lane (which runs into Limeburner's Lane) just off London's Farringdon Street, is named after the trade in coal brought to London by sea. Sea-coal was also often used to differentiate it from charcoal, the other major fuel of the day.

Re-imagining of 17th C glasshouse at Shinrone, County Offaly, ROI

The production of glass requires technical knowledge of many different processes and many different materials. It also requires the ability to heat a furnace to high temperatures and to be able to handle molten glass through all its stages before it becomes a useable product. In short, it has to be learnt from someone with experience and someone willing to impart their knowledge. Glassmaking was therefore a rare skill and one that glassmakers wanted to keep to themselves and their families and relinquish to others rarely, if at all. Whether it was a question of supply or demand – probably a bit of both – until the mid-1500s, glass was not in everyday use. The things that we see glass used for most commonly today, mainly containers for storing and selling a wide variety of products, drinking vessels, windows and lenses, simply did not exist. Goods of all types were not packaged, but bought loose, weighed or measured out in the street or market, and taken home in containers, baskets and panniers that the customer had brought with them. The 'grossers' (grocers) bought their trade-wares by the gross – 12 dozen – and sold them to the public in ones and twos. Drinking vessels used in homes and in taverns and ale-houses were made from much more durable materials than glass: pewter, animal horn or pottery of various types. Windows, if they existed at all, were generally unglazed (although panes of thin animal horn were known to have been used) and closed when they needed to be with covers of cloth or leather or with crude wooden shutters. Early houses appear to have had no openings at all and such light as there was came in through their doors. Glass therefore was in limited use.

Apothecaries and medical practitioners – one hesitates to call them doctors – might store various unguents, salves or precious liquids in glass phials; jewellery might be made using coloured glass beads or glass-type glazes; and stained glass was in use in ecclesiastical buildings. None of these uses, however, touched ordinary people as they went about their daily lives. Glass was also much more fragile than it is today and it wasn't suitable for serving or for drinking out of.

CHAPTER 4

Window glass

Probably one of the first uses of glass that showed what a suitable material it could be for the ordinary person was for windows. Although the Romans are known to have used small panes of glass to let light into their buildings, and its use in ecclesiastical buildings has already been noted, it wasn't until the invention and building of chimneys into houses that people appreciated that glazed windows would make their accommodation less draughty, warmer and therefore more comfortable. All the while that houses relied upon a central hearth for both heating and cooking, with the smoke finding its way up and out through thatched roofs, sealing up windows with glass was undoubtedly a disadvantage as the fire

Crown glass being spun by a glassblower

relied upon a constant movement of air to draw properly and for the smoke to escape. The Romans did have a form of central heating and water warming which involved the use of flues and chimneys in their houses and villas, and their bakeries are also said to have had chimneys to allow the smoke and heat to escape, but these techniques do not appear to have been adopted by the Anglo-Saxons when the Romans left. It wasn't really until the end of the twelfth century that buildings with fireplaces and chimneys started to be built, in the process allowing there to be internal floors and separate rooms much as we have today. Then, the requirement for windows which allowed in the light and kept out the wind and rain grew, and crude window glass began to be made.

For window glass, which of course is not hollow and is required to be flat, thin and as uniform as possible, there were two methods of making it. The early method was for the glassmaker to get a gather of molten glass onto the blowing rod, blow it into a small bubble and then, after transferring it to a pontil, spin it in front of a heat source so that the bubble flattened out into a large flat disc, known as a 'crown'. In experienced hands, this disc could have a diameter as large as 1.50 m (5 feet) and, after annealing and cooling, could be cut into small panes for use in 'lead-lights' i.e. windows constructed with lead in between the panes such as you find in stained glass windows. These small panes, known as 'quarries', were often diamond shape, which added strength to the window, and made more use of the edges of the disc which, of course, were curved. At the centre of the crown was the place where the pontil iron had been attached and this is known as the 'bull's eye' and is often a feature of these windows. Window glass produced in this manner was of different thicknesses, the glass nearer the centre of the crown being thicker, and glaziers would usually try and keep the thicker glass for the bottom panes of a window so that the window was not top-heavy.

As both raw materials and techniques improved, window glass began to be produced by being blown into a cylinder with parallel sides, which then had its top and bottom removed and was cut down one side. It was then reheated, unrolled and flattened out into a sheet. This got over the problem of the differences of thickness in crown glass, plus the wastage at the edge and centre of the circular crown. This technique, known as the 'muff' technique (after the shape of the cylinder), produced what was called 'broad glass' and became the standard glass for windows. This technique was further refined when hinged moulds were used into which the glass could initially be blown allowing for more uniformity of size and thickness. For the average person, window glass became the most important glass product. Once installed, it was less likely to break (than say a bottle or other container) and made their houses much more comfortable places to live in. The invention of the sash window and helpful changes in taxation on glass, further increased the demand for flat and even glass. The invention of plate glass in the late 1600s, made by casting glass onto a metal table and then grinding it flat on both sides, fuelled a demand by the wealthy for distortion-free mirrors and carriage windows and, if they could afford it, window glass. This further industrialised the glassmaking industry.

Glassmaking develops

In the Middle Ages, between the middle of the thirteenth century and the end of the sixteenth century (1250–1600), the main problems associated with glassmaking were problems which befell many other sectors of trade. Apart from relatively few main roads (and in many instances and at certain times of the year 'roads' is probably the wrong word for what they were), transport for both raw materials in, and finished goods out, was problematic. The centre of glass production in Britain at this time was the Weald of what is today Kent and East and West Sussex, then one of the most wooded parts of the country. Fuel, in the shape of wood, dried and cut to the right length and girth to fit the furnaces, and not requiring further splitting, was the principal requirement of glassmaking and was needed in much larger volumes than the actual materials for making the glass: sand, lime, flux and cullet. Wood was sourced

CHAPTER 4

from managed coppice woodlands; beech wood was preferred, although sweet chestnut was also used and coppiced, i.e. cut back to the 'stool' which would grow new shoots and eventually, after 12 to 14 years, be ready for cutting again. Special 'pot-clay' for the crucibles was also required and this was typically brought from more distant sources, although, of course, quantities were smaller than the volumes of fuel needed. Merret, in *The Art of Glass*, says that 'Purbeck in the Isle of Wight' has good crucible clay: 'the very same which makes Tobacco pipes'. Whether this should be 'Purbeck' or the 'Isle of Wight' is not known. Maybe both,

Coppiced lime trees

as both places have good pipe-clay. Merret also says that good pot-clay came from Nonsuch, a village near Epsom in Surrey, and from Worcestershire which, when 'mixed together make the best pots'.

Glassmaking families, many of them originating from what is today Belgium and Holland, migrated to the Weald and set up furnaces and started to produce a selection of glass products: crude window glass, small bottles and phials, bowls, cups and 'urinals' for the inspection of urine by the medical practitioners of the day. This *waldglas* was made using river sand as the silica source; wood or plant ash, especially from bracken, as the flux; and often colourants in the form of metals such as iron, copper or tin. Sand could have iron in it which also added colour. *Waldglas* in its natural state, i.e. without a colourant, varied from a very light green – just about suitable for windows and objects that need to be transparent to a dark green more suitable for bottles and containers where transparency was not required. The techniques to make (almost completely) clear glass were known, but carefully guarded, by the Venetians on the island of Murano, where the glassmakers lived in seclusion enjoying special privileges, but were not allowed to travel and impart their secrets. The technique of using crushed flint as a source of silica and adding lead oxide to produce a very clear, strong glass with a sparkling quality – ideal for cutting, decorating and engraving – was not discovered in Britain until 1677 when George Ravenscroft perfected its manufacture. It is often known (incorrectly) as 'lead crystal' glass after the Venetian *cristallo* glass, which used manganese oxide and no lead, but which is so named because of its resemblance to objects carved from rock crystal.

In 1549, Edward VI (then aged twelve) invited eight glassworkers from Murano, together with some Dutchmen, to come to England and set up a glassworks, which they did near today's Belsize Park in north London. The Venetian authorities took a dim view of this, as it was an infringement of the men's terms and conditions, and tried to get them back. For their own safety they were imprisoned in the Tower of London and eventually all but two, an Italian, Guiseppe Casselari, and a Dutchman, Thomas Cavato, returned, and attempts to make *cristallo* glasses ended, thus securing – for a while – the Venetians' monopoly over this much prized glassware.

In 1567, Jean Carré, a noted Huguenot glassmaker, French by birth, from Arras, but who had learnt his trade in Antwerp making window glass, moved to England, and together with his business partner Anthony Becku, was granted a monopoly on the production of window glass in 'the Lorraine or Norman' manner for a period of 21 years, so long as they trained local craftsmen in the art of glassmaking. Carré and Becku set up three furnaces on the West Sussex–Surrey border between Horsham and Hazlemere: two at Fernfold and one at Alford. Soon after setting up their venture the partners fell out and parted company, although Carré carried on in business and the glassworks are said to have flourished after he died in 1572. This part of the country became a centre of glassmaking, mainly window glass, and the Wisborough Green parish records contain entries for 26 French families between 1567 and 1617. (After that the entries abruptly end, owing to the 1615 Royal Proclamation which ended the use of wood for heating glass furnaces.) The names Henzey, Tyzacke and Bungards, all of Flemish descent, litter the world of English glassmaking over the centuries to come.

Carré also applied for a licence to make 'Venice glasses', i.e. made from clear *cristallo* glass, but he was turned down. Undeterred, he started a glasshouse, probably at Crutched Friars in the City of London, just south of today's Fenchurch Street Station, and employed Casselari, plus another seven glassworkers from Murano. On Carré's death in 1572, his brother-in-law, Peter Campe, and his son, John Baptist, tried to keep the glassworks going, but gave up and it was taken over by Giacomo Verzelini, a Venetian who had left Venice in the 1550s and had been living and possibly working in Antwerp, where his brother Nicholas was a glassmaker. Verzelini was a slightly mysterious character, changing his name several times, marrying well, working as a broker at times, and was, by all accounts, a well-regarded surgeon who was sent for at the birth of Robert Sydney, the future Earl of Leicester, when he was born in 1595. Between 1572 and 1574, Verzelini set up a glass furnace in Crutched Friars. In 1574, he was granted a 'patent of monopoly' for 21 years for the sole manufacture of drinking glasses, which were in turn, prohibited from importation.

It was granted on the same terms as Carré's, with the proviso that he should employ skilled Venetians and train locals in the art and science of making *cristallo* glass. Within a few years, the area around the glassworks became quite thickly populated with both Italians and French families, not only glassworkers, but also traders, silk merchants, herring exporters and musicians.

On 4 September 1575 the Crutched Friars glasshouse caught fire and whilst the 'forty thousand billets of wood' he had stored there didn't catch fire, the damage was bad enough and Verzelini moved temporarily to a furnace near Newgate whilst his original glasshouse was rebuilt. He moved back there in 1579 and he continued to run it until he retired in 1592. Glassmaking had made Verzelini a rich man and he retired to a substantial estate in Downe in Kent. Verzelini died in 1606 and he and his wife are buried in St Mary's, Downe, the same church as Charles Darwin and his family are buried. Verzelini's business was continued by his sons, but for whatever reason, it did not flourish. When the patent came to an end in 1595, it was given to Sir Jerome Bowes, Elizabeth I's ambassador to Russia as a reward for his loyal service. Bowes was awarded it for making drinking glasses in England and Ireland for twelve years. The Verzelinis attempted to wrest the patent back from Bowes, but did not succeed and by the end of 1595, their glasshouse ceased production.

One of the first references to a coal-fired furnace in England is contained in a letter written on 21 July 1587 by Sir Philip Sydney to his friend, Sir Robert Dyer, Earl of Essex, who was then imprisoned in Winchester House, London, which also had a glasshouse adjacent to it. In his letter, Sydney says:

> Add hereunto that very lately, by a wind-furnace, greene glass for windows is made as well by pit-coale at Winchester House in Southwarke as it is done in other places with much wast and consuming of infinite stores of billets [cut timber] and other wood-feull.

Here we have proof that new types of furnace were being constructed which burnt coal, and where the use of 'wind', i.e. a draught, was an important feature. The glasshouse at Winchester House is doubly interesting for this story, as 53 years later (1642), the prison at Winchester House is where Sir Kenelm Digby was held and where he carried out some of his glassmaking experiments. For how long there remained a glasshouse there is not known.

Other glassmaking patents followed in the next few years and the industry started to expand, but it soon became clear that fuel was the limiting factor and that the woods and forests in southern England were being seriously depleted for the benefit of ironmasters and glassmakers. Given the difficulty of obtaining wood for burning, some glassmakers started to investigate the use of coal which meant a move, away from the traditional glassmaking areas – Kent, East and West Sussex and Surrey – to places such as the Forest of Dean, where surface and shallow-mined coal was available. It was at this juncture, as the Elizabethan era came to an end, that glassmaking changed forever.

HMS Prince Royal, built in 1610, surrendered to the Dutch in 1665

Chapter 5

Glassmaking – the change from timber to coal

The first legal move against the use of timber for glassmaking in Britain was in February 1585, in the form of a bill which was 'against the making of glasses by straungers and outlandyshe men within the realm and for the preserving of tymber and wood spoyled by glasshouses'. The change from a peripatetic, wandering industry, always chasing the fuel, to one where the glassworks remained static and the fuel came to it, happened gradually. In the late 1500s and early 1600s, there is evidence that glassmakers were leaving the Weald, where the supply of wood was becoming scarcer, and gravitating to those parts of the country where coal appeared on (or not deeply below) the surface and could be mined without the need for deep shafts and long tunnels. Remember, at this time there were no mechanical pumps for removing water from mines, and the first successful steam engine for pumping wasn't in use in Britain until the very end of the seventeenth century. Surface coal was present in the Forest of Dean in Gloucestershire, conveniently near the River Severn for the transport in of raw materials and transport out of finished goods, as well as in parts of Staffordshire and Worcestershire.

Suitable clay for the construction of the crucibles was also extremely important for successful glassmaking, especially for coal-fired furnaces with their higher temperatures. A crack in, and eventual failure of, a clay crucible filled with valuable molten glass was both dangerous and expensive. There is plenty of evidence to show two things: that good pot-clay was transported to centres of glass production where no suitable clay was available locally, e.g. London; and that glassmaking (and indeed other trades that required clay for crucibles and retorts) centred on places which had mines nearby that yielded good pot-clay, e.g. Stourbridge. In some cases, pot-clay was found alongside coal and in these mines, the clay was as valuable a commodity as the coal itself. Lord Dudley (and his illegitimate son Dud Dudley) claimed to have been making glass using coal on his estate at Greensforge to the north of Stourbridge

The Knight Who Invented Champagne

Coal-fired *Verrerie Anglaise* from *l'Encyclopédie de Diderot et d'Allembert* published between 1751 and 1765

as early as 1613 and this area became an early centre for glassmaking, with furnaces situated in the surrounding countryside where both coal and suitable clay was available.[1] It was one of the first places where Huguenot glassmakers settled when glassmaking shifted from being wood-based to coal-based and, by 1861, over 1,000 residents were employed in glass production. To begin with, window glass and bottles were the main products, but the area eventually became known for its fine tableware and decorative glass. It was well-served by the Stourbridge Canal, completed in 1779, which (eventually) connects to the Severn just north of Kidderminster. The lock in the centre of Stourbridge is called Blower's Lock.

Influenced by glassmakers from Europe, where furnace design appears to have been more advanced, possibly because in some places coal had already superseded wood, English furnaces became larger, with winged extensions containing both fritting and annealing sections, thus helping make the whole process more efficient and certainly more economical in the use of valuable fuel. Obviously, the hotter temperatures achieved with coal, allowed for these side extensions. The main

---
[1] Dud Dudley also experimented with making coke from coal in order to smelt iron.

difference in these 'modern' furnaces was that as the heat source was in the centre of the structure (whereas with timber it was at either end of the fire-trench), the heat was more intense, smaller in size and less fuel was consumed. The deep fire-trench also brought air to the fire and allowed easier removal of the ash and clinker. All of these refinements allowed for the construction of narrower flues which provided extra draught which helped carry the heat and distribute it throughout the whole furnace and carry the smoke and fumes away from the working areas. Merret in *The Art of Glass* (written in 1611) states that 'Crystal furnaces are deep two feet' (he actually meant crucibles, not furnaces) and about 'twenty inches broad at the top' and can hold 'three to four hundredweight of Metall' (150-200 kg). Towards the end of the 1600s, furnaces had grown into large brick-built conical structures with tall chimneys. These buildings were known as 'glass cones' and typically had multiple crucibles each holding up to 500 kg of molten glass.

However, making glass with coal wasn't just a case of substituting one fuel for another. Coal has a much higher calorific value than air-dried wood (around three-and-a-half to four-and-a-half times higher per cubic metre[2]) and requires more draught to keep it burning and to distribute the heat around the furnace. Coal also produced sulphurous fumes which had two consequences. These fumes not only coloured the glass and made it darker, something those that used glass (for urinals for instance, where clarity is required) did not like, but also of course made working conditions difficult for the glassblowers and their colleagues. To overcome both of these problems, crucibles were covered and the kilns redesigned so that additional draught took the fumes well away from the working areas. In time however, glassmakers took against covered crucibles as they couldn't see what was going on, and it made the process of gathering the molten glass slower (and therefore more costly). They therefore put up with the darker colour of the glass and, for bottles, black glass became quite fashionable and was seen as more modern than the old, light green *waldglas*.

Glassmaking patents

The year 1610 was a notable one in the history and development of glassmaking in Great Britain. On 28 July, James I granted to Sir William Slingsby,[3] an admiral and 'Carver to the Queen', who was on the 1596 Cádiz raid (as was Mansell – see later),

---

[2] Forest Research: air-dried wood 1,400–2,000 kWh per m^3, compared to house coal at 6,400–7,300 kWh per m^3.
[3] Three others were also named on the patent: Andrew Palmer, Edward Wolverston and Robert Clayton.

a patent allowing him to 'erect furnaces to make bell-metal, glass etc with sea and pit coal for 21 years.' Whilst this patent didn't prohibit the making of glass using wood, and indeed, under the 'Case of Monopolies', which had been brought in 1602 about the sole right to produce playing cards, single monopolies covering all manufacturing of a certain object or class of objects were deemed illegal. However, the need for timber for the Admiralty to re-equip and rebuild the navy meant that many of the forests came under royal protection and there was, for all intents and purposes, a ban on using wood for glassmaking (and indeed other industries that needed great heat).[4] Timber was the main building material and also used for making barrels in very large quantities, used for transporting and storing a very wide range of goods.

In 1611, only a year after Slingsby's patent had been granted, Sir Edward Zouche, who had established the Falcon Stairs Glass Works in Southwark,[5] together with three partners: Sir Bevis Thelwell, Thomas Percival, and Thomas Mefflyn[6] were granted an 'absolute licence' to 'melt and make all manner of glasses with sea-coal, pit-coal, sucash[7] or any other fuel whatsoever, not being wood.' This was contested (one assumes by Slingsby) but on 4 March 1614, a second patent to Zouche was confirmed 'setting aside the rights of all the previous patentees' and forbidding the use of wood for glassmaking. As can be imagined, this caused considerable problems, with glassmakers operating under licence from Slingsby suddenly finding that they had new masters to deal with. Mefflyn, having died, was replaced by Robert Kellaway. For this all-encompassing 21-year licence, Zouche and partners had to pay £1,000 a year rent. However, the licence forbad: all others to make glass with wood; the importation of foreign glass; and retail glass-sellers to make contact with foreign glassmakers. (The patent was called into question by Parliament in 1614, 1621 and 1624, but was confirmed each time.)

Whilst Zouche and his partners had secured the monopoly on making glass with coal, making glass using wood wasn't totally banned and they set about changing this. It was at this stage that additional partners were sought and the Earl of Montgomery, Sir Thomas Howard and Sir Robert Mansell became partners, with Mansell assuming control and bringing his energy and expertise to the party. On 19 January 1615, the Slingsby patent was revoked and a third one, granted for the 'exclusive privilege to

---

[4] One source estimates that it took 4,000 mature oaks to build one ship-of-the-line (i.e. a warship) in 1801.
[5] The site of the glassworks is just downstream of Blackfriars Bridge by the riverside pub the Founder's Arms.
[6] Thelwell was a wealthy courtier and owner of the Minories Glasshouse; Percival was said to 'have invented the making of glass with coal'; and Mefflyn was the King's Glazier.
[7] The meaning of 'sucash' cannot be found.

make glass using sea-coal' was granted to Sir Edward Zouche's syndicate. Using their influence with the King and the king's advisors, mainly on the grounds of forest preservation, the partnership also managed to get an additional 'Proclamation Touching Glasses' issued on 23 May 1615 which prohibited the use of all wood and wood products (charcoal) in the glass industry. The proclamation stated:

> There hath been discovered and perfected a way and means to make Glasse with sea-cole, pit-coal and other Fewell, without any manner of wood, and that in as good perfection for beauty and use as formerly was made by wood

Taken together, with their patents and proclamations, Zouche and his fellow partners had secured a monopoly over the industry of glassmaking in England and ended the use of timber as a fuel (for glassmaking). Despite absolute monopolies being forbidden by law, something later reinforced by the 1625 Statute of Monopolies, warrants were issued to apprehend glassmakers trying to leave the country, glassware that had been made in non-Zouche-controlled glasshouses was seized and confiscated and fines were levied and paid. Mansell very soon realised that he needed to take complete charge of the business, so at some stage in 1615 he reorganised the partnership, paying each of the partners (including himself) an annuity of £200 a year. He was also paying the King £1,000 a year in royalties. In 1635 Mansell stated that he had invested £28,000 in the glass enterprise (a sum equating to many millions in today's values) and he was determined to protect his investment. He petitioned the Privy Council for 'letters of assistance' in order to pull down any glasshouses erected without his permission, and on 4 May 1618 he got an assurance from two glassmakers, 'Paull Vinion and Peter Comley', that they would 'not at any tyme hereafter directly or indirectly make any Glasse with wood'.

On 22 May 1623 Mansell was awarded the (very all-encompassing) sole patent for the 'Manufacture of Glass with Fuel, not being Timber or Wood'. This permitted Mansell to make:

> all manner of drinking glasses, broade glasses, windowe glasses, looking glasses, and all other kinde of glasses, glasses, bugles, bottles, vialls or vessells whatsoever made of glass of any fashion, stuffe, matter or metall whatsoever … with seascoale, pittcoale, or any other fewell whatsoever, not being tymber or wood.

With this patent, Mansell had secured his position as the pre-eminent person in the glass industry in England.

The Knight Who Invented Champagne

Sir Robert Mansell (1573-1656)

Sir Robert Mansell[8] was born in Margam, then in Glamorganshire (but today in Neath Port Talbot), and came from well-connected families. He was remotely related to Lord Charles Howard (Howard of Effingham), who was Lord Admiral for both Elizabeth I and James I and Mansell served under him as a captain on the 1596 raid on Cádiz, for which service he was knighted, aged only twenty-three. Some accounts say that Mansell served under Howard during the battle with, and defeat of, the Spanish Armada in 1588, when Mansell would have been 15, but this is by no means certain, although he was captain of a privateer in the West Indies in 1591, aged eighteen. Mansell served with distinction in various naval exploits – at one point capturing a carrack (a large sea-going cargo ship) full of valuable peppercorns – and after the 1602–1603 Battle of the Narrow Seas, he was made a Vice Admiral and then Treasurer of the Navy in 1604, taking over from Sir Fulke Greville who resigned (or more probably was forced to resign) after a significant amount of corruption in Navy finances was discovered. He was also very well connected and appears to have been a member of all the new trading companies formed to exploit the goods (and sadly the indigenous people) of newly discovered lands. He was a member of the Virginia Company in 1607, the Muscovy Company in 1608, the East India Company in 1609, the North West Passage Company in 1612, the Somerset Isle Company in 1615, and the Africa Company in 1618.

During the next ten years Mansell was engaged in various services for his king and country, not always very honourably, including in 1604 the hiring privately of a ship (which he part-owned) called the *Resistance* to carry ship's biscuit to England's fleet in Spain, but in fact which carried a commercial cargo of lead. The Admiralty was overcharged on the size of the vessel, duped on its cargo, and Mansell is said to have kept all the rigging which belonged to the Navy. Mansell then appears to have fallen out of favour with the Lord Admiral for questioning his powers to appoint a commission to look into 'abuses in state of the Navy' in which he knew that he (Mansell) would be implicated. The commission did eventually meet and investigate maladministration and corruption in the running of the Navy and, in the report that it produced in 1609, Mansell did not come out of it well. It showed that he had been taking backhanders and bribes from suppliers and it is said that he treated the 'Chatham Chest', the fund which was to provide retired Navy sailors with a pension,

---

[8] Mansell in his youth spelt his name Mansfeeld, but later in life it became Maunsell or Mansel. Mansell is the most common spelling used today.

CHAPTER 5

'as his own private bank account'. Mansell was sent to the Marshalsea prison for two weeks, but released and appears to have returned to his duties, although there were continuing problems and accusations over the next decade. He was friends with Henry, Prince of Wales, (who was to die of typhoid in 1612, aged only eighteen) who had an interest in reforming the Navy. Mansell and Henry toured the Chatham dockyards and inspected the Fleet in 1611 and they were fellow investors in the Virginia Company. Henry's untimely death didn't diminish Mansell's position, and in 1613 he staged a very successful mock sea-fight on the Thames to honour the wedding of the King's daughter, Princess Elizabeth. Mansell was married twice: firstly before 1600 to Elizabeth Bacon, whose father was Lord Keeper of the Great Seal and whose brother was Sir Francis Bacon; and secondly in 1617 to his mistress, Anne Roper,[9] one of the Queen's maids of honour. By neither wife did he have children. At his second wedding, the Queen (Anne of Denmark, wife of James I) is said to given the happy couple their wedding banquet at what was then called Denmark House (today's Somerset House in the Strand) and £10,000 as a wedding present.[10] Mansell was obviously very well connected and very well thought of at the time.

In 1618 the Navy Board was reformed, Mansell was forced out, although he managed to sell the office of Treasurer of the Navy and became 'Vice-Admiral of England', a more or less ceremonial title. In 1620 it appears he was back in favour, as he was appointed to lead an expedition to Algiers to rescue some English sailors who had been captured by pirates and sold into slavery. The expedition was dogged by problems – poor ships, ill health, lack of supplies, lack of gunpowder – and he returned with only a few of the captives, leaving many others behind. The Duke of Buckingham, after the king the most powerful man in the country, took against Mansell after this voyage, and by 1621 or 1622 Mansell's naval career was at an end.[11] However, this did not unduly worry him as he had other interests to pursue. Since 1601, despite his naval duties, Mansell had been a Member of Parliament for various constituencies: Kings Lynn (1601), Carmarthen (1603), Carmarthenshire (1604–1614). He also served as MP for Glamorganshire (1624–1625 and 1628) and Lostwithiel in Cornwall (1626). His name also appears on the Second Virginia

---

[9] There is some confusion about Mansell's second wife. In some sources she is called Ann(e) and in others Elizabeth. Some sources even mention a third wife.
[10] Page 77, *The Development of English Glassmaking 1560–1640*, Eleanor S. Godfrey.
[11] The Dictionary of Welsh Biography states that: 'contemporary naval papers suggest that Mansell's appointment as vice-admiral in 1618 was not in fact a promotion, but rather a deliberate removal to a less influential position because of his dishonest administration [of the Navy] from 1604 to 1618. The charges brought against him seem to have been well founded, but he remained secure in the favour of the king despite his inglorious expedition to Algiers in 1621.'

Charter of 1609, which was a list of investors who were attempting to establish a colony in North America.

In 1642 it was suggested to Charles I that Mansell be given command of the fleet, but others thought he was too old and nothing came of it. At times he was called an 'indifferent seaman' and a 'dishonest administrator' (of the Navy) but despite this, he had a long and diverse career. He died in 1656 at the age of seventy-nine.

Mansell and glassmaking

As we have seen, Mansell had been engaged in glassmaking since 1615, the year he joined Zouche's partnership which had a monopoly on the manufacture of glass in England. Then, having bought out his fellow partners, he set about establishing glassworks in various parts of the country, either under his own command and ownership or by others under licence.[12] Glassworks were established at Wollaston in Nottinghamshire, at Kimmeridge (see below), at Milford Haven in Pembrokeshire, Newcastle-upon-Tyne and several in London: Broad Street, Lambeth and Southwark. However, not all were successful and most of the above (except for Broad Street) eventually closed down.

The Kimmeridge glassworks was established under licence from Mansell by Abraham Bigo, a glassmaker, and Sir William Clavell, a landowner and financier, in 1618. Kimmeridge had (and still has) oil shale which was used to fire the furnaces, which halved the cost of the fuel when compared to timber. Excavations on the site of the old glassworks have unearthed significant numbers of fragments of glass bottles and flasks, many more than would typically be found at a *waldglas* site, suggesting that bottles were an important part of the output. The fragments of bottles recovered are made of thick glass (up to 9 mm (about a third of an inch)), and have long necks with a thick 'string-rim' around the top of the bottles. Phelps, writing in a 2010 English Heritage report, states that 'the bottle fragments point to the existence of "English bottles" before 1623 and are evidence of the Bigo family's involvement in the production of early wine bottles'. Clavell and Bigo, in contravention of the licence Mansell had granted them, started to send glass to London. Mansell took action against Clavell (the Bigo family had sensibly moved to Ireland to continue glass production there[13]) and as a result, on 23 August 1623,

---

[12] The King was amused by Mansell's involvement in glass saying why Mansell 'being a seaman, whereby he got so much honour, should fall from water to tamper with fire, which are two contrary elements'.

[13] A glasshouse at Shinrone in County Offaly in Ireland, which was associated with the Bigo family, has

Clavell was committed to the Marshalsea prison – the fate of many a debtor at the time – and the glassworks was closed down. After being released on a petition of ill health, he reopened the works and promptly found himself back in the Marshalsea. The final act took place in 1626 when he agreed to forswear glassmaking in return for an absolute pardon for any crimes he may have committed, including witchcraft! This was the end of Clavell's glassmaking enterprise at Kimmeridge. His debt now stood at £20,000, ten times the previous level, and it took three generations of scrimping and saving for the family to pay it off.

Mansell's Newcastle-upon-Tyne glassworks, established in 1618, was successful and made glass and shipped coal to his other glassworks. However, it wasn't all plain sailing for Mansell: disputes of one type or another continued for many years and there were several complaints made against the quality of Mansell's glass. The Glaziers' Company said that it was 'mostly unserviceable' and the noted architect, Inigo Jones, said it was 'mixed good and bad together, and is very thin in the middle'. All this arguing and complaining often went on when Mansell was away at sea, carrying out his naval duties. When this occurred, his wife, Lady Anne,[14] managed the business and fought his corner with great effect. Together, the Mansells managed to defend their interests, so much so that in the 1624 Statute of Monopolies, they are specifically excluded, by name, from the provisions of the Act.

Mansell's Newcastle glassworks evidently prospered, no doubt because the fuel (coal) was plentiful and shipping it in colliers[15] was a very cost-effective way of getting it to a glassworks. In 1623, Mansell had his patent cancelled by Parliament, but it was immediately renewed for another 15 years, and in 1634 renewed again for a further 21 years in exchange for an increase in his annual rent from £1,000 to £1,500. In 1624 he secured a 21-year lease on a mine to produce fireclay to make retorts, one of the most important items in a glassworks. In 1624, during the debate in Parliament on the Monopoly Bill, Mansell defended his interests, saying that he had established, and in some cases tried to establish but failed, glassworks in London – at Lambeth, Southwark and Broad Street – Kimmeridge on the Isle of Purbeck, Wollaston in

---

been excavated and fragments of English-style 'shaft and globe' bottles dating between 1620 and 1641 have been identified. It is the only glasshouse from this date in the British Isles to have part of its superstructure still standing.

[14] See previous footnote about the name of Mansell's second wife.

[15] The 'colliers' of this part of north-east England were such sea-worthy ships that the Admiralty purchased a Whitby-built one in 1765, then called the *Earl of Pembroke*, re-fitted it and re-named it HMB *Endeavour* and sent Captain James Cook on it to Tahiti to observe the transit of Venus and then travel east to try and find *Terra Australis Incognita* or what we today know as Australia and New Zealand. (The 'B' in HMB stands for 'bark' or 'barque'.)

Stourbridge Canal and the Red House Glass Cone

The Red House Glass Cone, Stourbridge was built around 1790 and made glass until it closed in 1936. It is now a glass museum.

Nottinghamshire, and Milford Haven. One of the era's more successful glasshouses was the Haughton Green Glasshouse in Denton, Manchester. Here, according to Ellis in his book on the Stourbridge and Dudley glassmakers, the du Houx family made glass under licence from Mansell between 1615 and 1653. The quite substantial glassworks was excavated between 1969 and 1973 and fragments of bottles, amongst other glassware, were discovered.

In his evidence, Mansell also said that he employed '40 sayle of ships' and '4,000 natives' to transport coal and glassmaking materials around the coast, and that he employed 500 people in the making of 'looking glasses' (mirrors). He also added that he was saving a huge amount of wood from being burnt. It is also said that he had an interest in a glasshouse at Newnham on Severn which was possibly engaged in the production of bottles and where it is possible that Sir Kenelm Digby, of whom much more later, was involved.

In 1630, Mansell secured an Order in Council restricting the importation of certain types of glass but, despite all these safeguards, in 1635 he complained that he had sunk £28,000 (a vast sum of money at today's value) into the whole business before he saw any return and that Scottish glassmakers enticed his men away and then he was forced to hire them back at inflated wages. In 1640 Mansell suffered further problems at his Newcastle-upon-Tyne glassworks when the Scots invaded England, beating the English Army at the Battle of Newburn (or Newburn Ford) and occupying the city. In order to pay the Scots off, Charles I recalled Parliament (the Long Parliament, so called because it met for 20 years[16]) which effectively led to the start of the Civil War two years later in 1642.

Mansell's complaints about others infringing his patents eventually led to an investigation by the Committee of Grievances and resulted in an order that Mansell hand back his patent and end his monopoly, which he did in 1642. In his evidence, Mansell told the House of Lords how he had reduced his prices: beer glass from 6 shillings to 4 shillings per dozen; ordinary wine glasses from 4 shillings to half a crown (2 shillings and 6 pence) per dozen; cristall beer glasses from Venice from 24 shillings to 10 shillings and 11 shillings per dozen; cristall wine glasses from Venice from 18 shillings to 7 shillings and 8 shillings per dozen; and window glass down to 22 shillings and 6 pence 'per case of 180 foot' of glass 'except for a small quantity made at Woolwich.' Mansell eventually got his Newcastle-upon-Tyne glassworks back and kept it running until he was unable to renew the lease in 1656, the year he died.[17]

---

[16] The reason it met for 20 years was because one of the first things it did was to pass a bill stipulating that it could only be dissolved by agreement of its members, and not by the monarch. The members didn't agree to dissolve it until after the Civil War and just before the Restoration of the Monarchy in 1660.

[17] The Oxford Dictionary of National Biography states that: 'The parish register of St Alfege, in East Greenwich, where [Mansell] lived, indicates that he was buried on 21 August 1652, so it is difficult to explain why licences to export horses were issued in his name in October 1655.'

Chapter 6
Glass bottles

Glass bottles of one sort or another had always been part of the glassmaker's output. Unlike earthenware or stoneware, glass did not need glazing internally to be watertight, and the contents and fill height (the level of liquid in the bottle) could be seen by all, assuming the glass was light enough in colour to be seen through. A glass bottle could also be made into an attractive shape that would grace the dinner table and enhance the attraction of the product inside. The drawback of glass was that it was fragile, so transporting bottles, whether empty or containing liquids, over rough roads was problematic and, once chipped or cracked, they were unusable (although they could be recycled). The internal volume of a bottle that was holding a liquid intended for sale also needed to be consistent, although a good glassmaker, by using the same weight of glass to start with, and by blowing a bottle to the same dimensions (which could be measured with callipers) could achieve very consistent results. Early bottles were therefore usually only used for serving from, like modern-day decanters, and for short-term storage and never used at the point of sale, as we do today.[1] This really

---

[1] Until 1860, when William Gladstone, then Chancellor of the Exchequer, changed the structure of Excise Duties on alcohol and then, in 1861, introduced the Single Bottle Act which allowed all shopkeepers to sell wine by the single bottle for consumption 'off the premises' – hence the term 'off-licence' – bottled

The Knight Who Invented Champagne

Reproduction shaft-and-globe bottle made for *Inside the Factory* TV programme about sparkling cider

didn't change much until the development of moulding techniques, which guaranteed consistency of size and internal volume, but this did not happen until the early 1800s.

Commonly consumed alcoholic beverages such as wine, ale, beer, cider and mead, whether being drunk in a private house, mansion or palace, or served in an ale-house or tavern, would arrive at the premises from the brewery, the cider-maker, or the vintner in barrels. These would then be stored in a suitable cellar and their contents drawn off from the barrel as needed into serving vessels, which might include directly into tankards or flagons if the barrels were stacked behind the bar, or into glass bottles. In grand houses, mansions and palaces, the task of bottling supplies of wine and other drinks for the table would fall to the 'bottler' (from the Old French *bouteiller*) – which is where today's word 'butler' comes from. Since at least the Norman Invasion of 1066, the 'Keeper of the King's Wines' was known as the King's Butler of England or *Pincerna Regis*. The King's Butler was also able to charge duty of 2 shillings per imperial ton of wine imported into England by 'merchant strangers', i.e. not British traders, a right which started in 1302 and did not stop until 1809.[2]

The development of bottles as containers for liquids undoubtedly progressed organically, moving from squat bottles made for holding precious liquids to something with a base and a neck – more like a typical wine bottle of today. Willy Van den Bossche in his magisterial *Antique Glass Bottles, their History and its Evolution* (2001) states that as early as the ninth century glassmakers in mainland Europe were making rudimentary bottles, and by the 1100s had developed the 'shaft and globe' type bottle by blowing a simple bladder and then swinging it in an arc up and over the glassblower's head to stretch and elongate the neck. It was found that the blend of

---

wine was a rarity and not available to the general public.

[2] The position of King's Butler was created in 1068 and was traditionally associated with the manor of Kenninghall in Norfolk. Since around the 1850s it has been claimed by the Duke of Norfolk, todays Earl Marshall, although this claim is disputed. The Chief Butler's main task was to arrange the Coronation Banquet. He was also known as the 'Taker of the King's Wines *captor vinorum*'. In London, the Chief Butler was also the City Coroner and Chamberlain.

materials in the 'metal' needed to be 'two-parts sand, one-part tree ash and some cullet' to achieve the right consistency of glass. Once the glassblower had achieved the final bottle shape, the blowing rod was detached from the mouth of the bottle (where the air from the glassblower's lungs had been introduced to create the void in the bottle) and a pontil iron attached at the bottom of the bottle with a blob of glass to act as adhesive. This allowed the glassmaker to cut the neck cleanly and put a string-rim (a glass collar) around the neck. This rim would serve two purposes: it would help prevent the mouth of the bottle from getting chipped and could be used to secure a stopper – typically a wooden peg with cloth wrapped around it – with string so that the liquid would not easily escape.[3] Once the neck had been finished, the bottom of the bottle was pushed up slightly, forming a base for the bottle to rest on and the pontil iron would be detached, leaving a rough mark. This created what we now call the 'punt' (although glassmakers call it a 'kick-up') which, on very old bottles, still contains the scar left where the pontil iron was originally attached.

When the first true glass bottle was made, one with a flat base upon which it could stand up and a neck which could take a stopper of whatever type, i.e. one that today we would recognise as a bottle, is open to debate. Bottles for holding liquids had certainly been in use since Roman times and quite probably earlier, but these were bottles for storage of precious liquids, for use by apothecaries and doctors as urinals for inspecting urine and as containers for their precious liquids. They were not bottles that would travel well, survive handling at their destinations and had a guaranteed volume making them suitable for selling products in. Before the advent of proper glass bottles, the word 'bottle' was often used to describe a ceramic or even leather bottle. Shakespeare, who lived from 1564 until 1616, and whose plays, sonnets and other writings contain almost one million words, used the word 'bottle' in relation to drink quite rarely: for wine 12 times, beer once, aqua vitae once, leather once and in stage directions twice. He also uses it in the sense of 'lost

---

[3] Corks do not appear to have been used in Britain before 1530 and the first British patent for a corkscrew wasn't granted until 1795.

Crude glass bottle neck with hand-applied 'string rim'

his bottle' but that's a different context. He uses the word 'cork' – in relation to drink – twice. This suggests that the words 'bottle of wine, beer, ale or mead' were not widely used by the general public for whom Shakespeare wrote his plays. In Neri's *The Art of Glass*, written in 1611 and at the time, the *only* book on glassmaking, the word 'bottle' only appears four times. However, two of these mentions are in an appendix listing 'instruments used in making green glass'. These include 'ferrets' which are 'those Irons which make the Ring at the mouth of Glass Bottles' and 'fascets' which are the 'Irons thrust into the bottle to carry them to anneal'.[4] It would seem therefore that bottles were by then part of the glassmaker's output.

Gervase Markham's *The English Housewife* published in 1615, discusses the brewing of 'Bottle Ale' saying that it should be: 'put it into round bottles with narrow mouths, and then stopping them close with corke'. He continues: 'be sure that the corkes be fast tied in with strong pack-thrid [pack-thread], for fear of rising out, or taking vent'. This book also suggests that bottles were used to hold liquids in the mid-1500s, but these were 'wanded' bottles, encased in wicker or leather containers, similar to the straw covering of old-style Chianti bottles. This was to protect the bottles against damage during transport or use. The bottles lacked a string-rim, so this meant that they were probably not sealed tight with wooden pegs wrapped in cloth or with corks, both of which would need tying down for safety. If they were only used for serving wine in ale-houses and taverns, then they maybe didn't need stoppers at all.

Roger Dumbrell, in his 1983 book *Understanding Antique Wine Bottles*, states that by 1577 glass bottles had become a sign of status. In 1589 two glassmakers, Hugh Miller and Acton Scott, petitioned the Queen 'for a Lease for term of Years, to make certain kinds of Glasse, namely Urynalls, bottels, bowles, cuppis to drinck in and such lyke' at

---

[4] Websters Dictionary, 1913 edition, says that a 'fascet (plural fascets) is a wire basket on the end of a rod to carry glass bottles, etc., to the annealing furnace'.

the Crutched Friars glassworks. This was for 'bottles and vessels', showing the importance of this type of product. Glass at this stage was made from the usual culprits: sand, lime, potash, alumina and oxide of iron – plus, of course, cullet. The sand and lime were the largest constituents by weight, followed by the cullet to help it melt, and the potash, alumina and iron oxide to give it workability, strength, and colour. As the use of bottles grew, so did the amount of broken glass – cullet – available to be collected and for some people it became their main trade, visiting houses, pubs and taverns where they knew glass was used, and selling it back to the glassworks. No doubt the servants in the houses, pubs and taverns exploited this trade to earn a little extra pocket money.

Most historians and writers suggest that it was in the 1630s, coinciding with the more widespread use of coal as a fuel for glass furnaces, that bottles for drinks started to appear regularly. Both Dumbrell and Van den Bossche suggest that the earliest 'shaft and globe' bottle, with a string-rim, dates from 1630 at the earliest, with Van den Bossche narrowing it down to 1632–1634. The example he has in his book from this date is of Belgian origin. These dates, of course, coincide exactly with when it is suggested that Digby had 'invented' the glass bottle. Dumbrell states that, between 1592 and 1620, there is very little evidence of wine bottles having been made in England. By 1633, however, they definitely were, as they appear in household accounts and inventories, and by 1634 they are listed in the London Port Books as export cargoes. By 1645, Oliver Cromwell thought the trade substantial enough to place a tax on glass, of all sorts. Some households got through enormous quantities of bottles in a year – some household accounts talk of 500 bottles being used – showing that they were something of a luxury and a status symbol. If you were rich enough to have your wine served in bottles, how rich would you have to be if you didn't bother to look after them?

The evidence therefore points to a sea-change in glassmaking (and therefore bottle-making) between 1610 and 1615, occasioned by the switch of fuel from cooler-burning wood to hotter-burning coal. This led to the development of better glass and, with it, bottles that were both practical and durable. Within a few years, glassmakers, together with landowners who owned coal mines and had the capital to fund the building of permanent and more substantial glassworks, could see that bottles were going to be one of their major products in the future. Mansell takes over the Zouche glassmaking consortium in 1618, just as glass bottle-making was taking off and he takes it to a new level. Digby, born in 1603, was only 15 at this time and had adventures of his own to complete. But by 1629, when he returns from his voyage of discovery and starts experimenting with glassmaking, he is 26 and ready to get involved. It is at this period that he has to meet Mansell, take an interest in the Newnham on Severn glassworks and 'invent' the glass bottle.

Glass bottles in the wine, beer and cider industries

Once bottles were being made that were robust enough to withstand the demands of handling and transport and with a capacity that was pretty much the same in each bottle, the wine, beer and cider industries started to use them. Ale and beer could be made every week, all year round, so long as grain (usually barley), hops (required for beer but not ale), yeast and water were available. This made it a product less likely to require bottles, as beer was generally brewed one week and drunk the next. Beer and ale are, of course, more suitable for mass-production in industrial-sized breweries than cider or wine, making them cheaper to produce. At this time, beer or ale would be delivered from the brewery, drawn direct from the barrel once it had settled and clarified, and served to the customers. In all probability, many taverns and ale-houses had their own brewhouses so they didn't even need to transport it far.

One of the first entries in Pepys' diary, the one for Monday 2 January 1659/60, mentions that 'old East' had brought him a 'dozen bottles of sack'. East was a servant of Pepys' boss, the Earl of Sandwich. In the same diary entry, Pepys says that Mr Sheply, another Sandwich employee, who was engaged in the 'drawing of sack in the wine cellar' had been told to give him the wine as a gift. Another entry (of 1 August 1660) talks of going to a 'bottle beer house in the Strand' making a distinction between this establishment and an 'ale-house'. It is more likely that the beer in the 'bottle beer house' was still drawn directly from the barrel, but rather than being drawn into pint (or quart) pots made of pewter, pottery or even leather and then given to the customers, it was drawn into what then would have been fairly unusual clear glass bottles, probably considered a fashionable novelty, and placed on the table. It would be similar to a wine bar today, where instead of serving the wine in the bottles it came in, the wine is decanted into a classy decanter before serving. The routine bottling of ales and beer does not appear to have become commonplace until after 1845 when the glass tax was removed and mould-made bottles were developed and a screw-top beer bottle was invented (1879). Until then, some beers were bottled in (very heavy) pottery bottles, and corks, expensive to buy and laborious to insert, were used to seal them.

Once reliable bottles were produced, cider was probably the largest user of bottles in Britain. Unlike wine, cider was made from native fruit, could be made over quite a long period (apples can be stored without refrigeration for around four months after harvest) and was much more the drink of the common man who frequented the alehouses, where most alcohol was then consumed. For cider, however, bottling appears to have been quite frequent in the 1660s and on 10 December 1662, in a paper read by Henry Oldenburg, the first Secretary to the Royal Society, to its members gathered in Gresham College, the Rev. John Beale of 'Yeavil in Somersetshire' stated that:

> Bottleing is the next improver, and proper for Cider; some put two or three Raisins into every Bottle, which is to seek aid from the Vine. Here in Somersetshire I have seen as much as a Wal-nut of Sugar, not without cause, used for this Country Cider.[5]

This would appear to be a recipe to guarantee sparkling cider. Beale also stated that cider was much better when bottled and kept cool, and that the 'King, Nobility and Gentry did prefer it before the best wines those parts afforded', 'those parts' being Gloucestershire and Worcestershire. From contemporary accounts it is clear that people believed (probably quite correctly) that bottling improved cider. Another member of the Royal Society, Sir Paul Neile (also spelt Neil), read his paper a 'Discourse on Cider' to the Royal Society on 8 July 1663 and stated:

> But the remedy is, in case you be put to a necessity to use it, that you open every bottle after it hath been bottled about a week or so, and put into each a little piece of white Sugar, about the bigness of a Nutmeg, and this will set it into a little fermentation, and give it that briskness which otherwise it would have wanted.

On 22 July 1663, Captain Silas Taylor also presented a paper on cider to the Royal Society. In it, he describes bottling cider and keeping it in cool water which makes it:

---

[5] This was one week before Dr Christopher Merret read his now famous paper which mentioned adding sugar to still wine to make it sparkling.

drink quick and lively, it comes into the glass not pale or troubled, but bright yellow, with a speedy vanishing nittiness[6] (as the Vintners call it) which evaporates with a sparkling and whizzing noise.

Cider appears to have been a popular topic with the Royal Society in its early days and in December 1663 John Evelyn was instructed to gather together all the different discourses and observations on cider written by the Reverend Beale, Sir Paul Neile, John Newburgh, Dr Smith, and Captain Taylor and publish them. This he did in 1664 as a section called 'Pomona', published on the express orders of the Royal Society, which was added to the end of *Sylva*, his book on trees and the timber industry.

Pepys gives us several instructive examples of the use of bottles in the 1660s. On Friday 23 October 1663 he writes: 'Thence to Mr. Rawlinson's at the Mitre and saw some of my new bottles made, with my crest upon them, filled with wine, about five or six dozen.' Whether these were the first bottles Pepys bought with his crest on, who knows, but he thought it worth making a diary entry about them. Pepys's diaries are a great source of information about daily life in London in the 1660s (they run from 1660 until 1669) and are especially good about eating and drinking. He visits 153 separate taverns, mentions wine 345 times and there are entries relating to 20 different alcoholic drinks. He also liked fine wines and, as has already been said, he was one of the very few writers of the time to mention a wine by the name of an actual château: Ho Bryan (Haut-Brion). Pepys was friends with and/or knew all of the characters in this tale: Charles II, Mansell, Digby, Merret and Winter. Pepys was born in 1633 (and died in 1703), so was of a younger generation, but was made a member of the Royal Society and was its President in 1684–1686. He could have undoubtedly told us much about glass bottles, their invention and their manufacture.

The Ashmolean Museum in Oxford has a unique collection of very early 'globe and shaft' wine bottles which have been excavated from five different taverns in the city. Knowing who the licensees were at the time, the museum is able to both date the bottles and know, by the seals on the sides of the bottles – the way of marking whose bottles they were – that they belonged to the landlord and not his or her customers. The earliest bottle in their collection is from the Three Tuns tavern and belonged to the first landlord, Humphrey Bodicott, who took the tavern over in 1639. His daughter Judith took the tavern over in about 1658, after her father's death, so the bottle dates with some certainty from this period. When his daughter died in 1666, her will lists

---

[6] The Oxford English Dictionary describes 'nittiness' as being 'full of small air bubbles (referring to wine)'.

wine worth £191 14s. 0d., equating to 600 imperial gallons (2,727 litres) of wine, the equivalent of around 3,636 of today's 75 cl wine bottles. Another bottle, this time from a pub called The Salutation in Cornmarket Street (then called the High Street) dates from between 1647 and 1670 when Thomas Wood was the licensee. In 1651 Wood took the lease on a house at No. 104 High Street and also obtained a licence to sell wine from the premises. This house belonged to Oriel College and behind it and No. 105 was a tennis-court for playing 'real tennis' i.e. not lawn-tennis. This was run by Wood together with the tavern and his wine bottles bore a seal showing two tennis players. No wine bills from this time have been found in Oxford archives, but entries in the account books for All Souls College show that Wood supplied the College with wine from 1652 to 1663. One very fine example of an Oxford bottle with a seal is from the Crown Tavern, then a tavern at No. 3 Cornmarket Street (not today's Crown pub further down the street on the opposite side) which was then owned by New College. This bottle, now in the Corning Museum of Glass in Upstate New York, has a seal showing a crown with OX and ON (for Oxford) either side, the date 1678 at the bottom, and a combination of initials standing for the names of the licensees, William and his daughter, Anne Morrell. It would seem, therefore, that taverns took delivery of the wine, bottled it in their own branded bottles, and no doubt charged a deposit on the bottle so that their customers would return them clean and empty. Only the largest houses and, of course – it being Oxford – the colleges, would have had use for a whole barrel's worth of wine at one time.

Seal from a bottle found at the site of the Three Tuns Inn, Oxford

John Worlidge in his book *Vinetum Brittanicum*, or a Treatise on Cider, first published in 1678 goes to some length to discuss corks and how to prepare them for bottling, but says that he prefers to grind the necks of bottles so that they are perfectly smooth and then fit ground glass 'stopples' to each bottle. He does however point out that these 'stopples' fit so tightly that the 'Liquors are apt to force the Bottles', i.e. break them, and care must be taken. The book also contains lots of references to bottles, corks and sugar and also has the first description of storing bottles horizontally with their necks down in a wooden rack – perhaps the precursor of today's *pupitre*? In a book called *Vinetum Angliæ*, 'a new and easy way to make wine of English grapes and other fruit: equal to that of France, Spain, &c. with their physical virtues' published in 1690 and

Bottle from The Crown, Oxford, 1678

written by someone we only know by the initials 'D.S.',[7] mention is made of 'adding little lumps of Loaf sugar to cider so that it may the better feed and keep'. D.S. then continues, when describing Perry making, by saying 'work it as the Cyder, and put in a few lumps of Loaf Sugar for it to feed on; and being well fined, and drawn off, it will drink brisk, and exceeding pleasant'.

By the end of the 1600s, there were 39 glasshouses in England producing bottles, of which 13 were in the cider-producing west of England. In May 1695, a tax of 1 shilling a dozen was introduced, which cut the number of bottles being produced, so much so that it was repealed in 1699. Production of bottles had ceased in both Newnham on Severn by 1715 and in Gloucester by 1741, but in Bristol and the surrounding area, it was flourishing and 'English glass bottles' were being shipped to both Europe and North America. By 1725, Bristol was the region's centre for glass, and making glass bottles using sand brought downriver from Newnham on Severn. The nearby village of Nailsea, where coal mining started in the 1500s, became a well-known centre of glassmaking, and by 1835 had one of the largest bottle-making facilities in Britain. The site of the glassworks is now covered with a Tesco car park, but there is a fine mosaic in the supermarket depicting the different forms of glass that used to be made there.

---

[7] The book was 'sold by G. Conyers, at the Gold Ring in Little Britain (Price One Shilling)'.

Chapter 7
Sir Kenelm Digby (1603-1665)

The crowd surged forwards and a small boy held his mother's hand tightly. He had also heard the horse's hooves and jeers and whistles of the crowd and waited in trepidation. A public hanging and disembowelling, especially of four of the traitors who had tried to blow up the King and Parliament, was not an everyday occurrence. It was Thursday 30 January 1606 and Mary Digby and her three-year-old son Kenelm (and her one-year-old son John) were waiting in the cold on Cheapside for the execution of her husband, and the boys' father, Sir Everard Digby, on the gallows set up at the west end of Old St Paul's Churchyard. As his father passed by, it is said that Kenelm said 'tata, tata'.

Sir Everard Digby, said to be 'the handsomest gentleman in England', was one of the infamous Guido Fawkes gang, known by history as the Gunpowder Plotters, who had tried, and narrowly failed, to blow up the Houses of Parliament on 5 November 1605. The nine surviving traitors (there were 10, but the ringleader, Robert Catesby, had been shot whilst resisting arrest) had been tried in Westminster Hall and, alone amongst the nine, Digby had pleaded guilty, which gave him the right to make a speech before he met his fate. He was also the first to be dealt with. He mounted the scaffold and addressing the large crowd said that even though he had broken the law, morally and – perhaps more importantly – in the eyes of his religion, he had committed no crime. He refused the attentions of a Protestant clergyman, speaking to himself in Latin, before saying goodbye to his friends and asked for God's forgiveness and protesting the innocence of Father Gerard and the Jesuits. Sir Francis Bacon, who attended the executions, said that as the hangman plucked out Digby's heart, saying 'here is the heart of a traitor', Digby had protested his innocence saying 'thou liest' as he died. What effect this early experience had upon the young Kenelm Digby is anyone's guess, but they must have been some of his very first memories.

Drawing, hanging and quartering,[1] the statutory penalty since 1352 for those guilty of high treason, was a gruesome affair. The victim was strapped to a wooden hurdle, usually backwards and upside down 'at the horse's tail', facing the rough street cobbles 'so that he should not pollute the common air', and drawn through the streets from the Tower of London to the place of execution. The hooded hangman then put a noose around the criminal's neck, strung him up until partially throttled, then lowered him down from the gallows. Then, as the criminal struggled to get his breath, his sexual organs were removed, he was then beheaded, disembowelled and quartered, the parts being distributed to the 'four corners of the realm' to be prominently displayed as a warning to others. Women convicted of high treason, because of considerations of public decency, were merely drawn through the streets and then burnt at the stake.[2] The Attorney General, Sir Edward Coke, who had prosecuted the plotters, had said at the end of the trial, that they 'should be drawn backwards, their genitals cut off and burnt before their eyes, their bowels and hearts removed, decapitated and dismembered' and the parts distributed so that they were 'prey to the fowls of the air'. Guido Fawkes, the plotter who was caught ready to set light to the gunpowder under Parliament, thwarted the plans of the hangman by leaping from the gallows' platform with the noose around his neck, dying as he fell.

Kenelm Digby, the eldest son of Everard and Mary Digby, was born on 11 July 1603[3] at the family home of Gayhurst[4] in Buckinghamshire. It was just over three

---

[1] Not 'hung, drawn and quartered'. The drawing refers to being drawn through the streets, not 'drawing' as one would draw out the intestines of a chicken or turkey before cooking.
[2] Although the Act of Parliament defining high treason remains on the United Kingdom's statute books, for a long period of the 19th century, the sentence of drawing, hanging, and quartering survived. But, after legal reforms, was changed to drawing, hanging until dead, and posthumous beheading and quartering, before being abolished in England in 1870. The last public beheading in England was of the Cato Street conspirators in 1820. The death penalty for treason was only abolished in 1998.
[3] Some references say 11 June, but John Aubrey in his *Brief Lives* refutes this, stating that it was Ben Johnson who wrote 'June' to rhyme with 'Scandaroon'.
[4] It had been called Gotehurst. During the Second World War, the house became 'Outstation Gayhurst', a satellite of Bletchley Park, which housed five of Alan Turing's vast 'bombes', responsible for decoding the German Enigma machine's messages.

months after the death of Elizabeth I and the accession of her second cousin, James I (James VI of Scotland) on 24 March 1603 to the throne of England. The Digby family were a long-established Catholic family with land in Buckinghamshire acquired through inheritance and marriage. As was common with most Catholics at the time, they had ostensibly converted to Protestantism in order to keep their wealth intact. Everard had at one time been a courtier at Elizabeth's court and was sufficiently well respected by the new king to be knighted on 24 April 1603 at Belvoir Castle, as the king progressed to London from Scotland. Four days later the king and Digby senior were both present for Queen Elizabeth's funeral in London.

The previous 70 years had been troublesome for Catholics and Protestants alike, and many had died on both sides of the religious divide brought about by Henry VIII's split from the Church of Rome in 1534. The Digby family had played their cards well, keeping in with the right people and, with a substantial income from their landholdings, lived a life of wealth and privilege. In 1605 they came under the influence of the notorious Jesuit priest John Gerard who had been active since the 1580s in helping Protestants convert to Catholicism, a service he performed for both Digby and his wife, despite the fact that she had been brought up a Protestant. Gerard had suffered torture on several occasions for his beliefs and had famously escaped from the Tower of London and lain low for eight years whilst still continuing his work. After the Gunpowder Plot, in which he was implicated, he escaped to France and then travelled to Rome where he died, in his bed, in 1637.

After his father's execution, young Digby inherited his father's estates and income, reputed to be around £3,000 a year – worth at least £1 million in today's purchasing power. The state attempted to wrest the properties from him, but his father had foreseen this and had 'entailed' the property so that it was not his when he died, making sure that it passed to future generations. Shortly after his father's execution Digby was placed with a Protestant family and brought up in that faith. One of his main tutors during this period was William Laud, then Dean of Gloucester, but who went on to became Archbishop of Canterbury in 1633.[5] In 1617 Digby's distant cousin,[6] Sir John Digby (later the first Earl of Bristol), who at the time was English ambassador to Spain, took him, aged fourteen, to Madrid for six months. One assumes that whilst he was there, he would have learnt at least a modicum of Spanish.

In 1618, aged only fifteen, Digby became an undergraduate at Gloucester Hall,

---

[5] Archbishop Laud, who advanced his career under the patronage of Charles I, was eventually accused of treason, found not guilty, but nonetheless executed in 1645.
[6] Most references say with certainty 'cousin'; some say 'uncle'.

Oxford University, one of the few of the University's colleges not to have a chapel and which was therefore able to recruit students who were Catholic or at least Catholic sympathisers. Here he came to the notice of the famous mathematician and astrologer Thomas Allen (or Alleyn) who remained a lifelong friend, bequeathing Digby his collection of 200 plus books on his death in 1632.[7] Digby stayed at Oxford for only two years, never graduating (then quite a common occurrence for Catholic-leaning undergraduates), and in 1620, aged seventeen, he was sent on a Grand Tour, as was quite usual in those days for young men of wealth and influence. He first went to Paris, where he met the famous mathematician Pierre de Fermat (he of the Last Theorem) but left there when plague broke out and travelled to Angers. There he met and impressed the forty-five-year-old Marie de' Medici, the feisty widow of Henry IV of France and at that time the Queen Mother to her son, Louis XIII, father of Louis XIV, the Sun King. It is said that Marie de' Medici took a distinct liking to Digby, so much so that he had to 'flee her royal attentions' and had it put about that he had died so he could make his escape.

Digby then travelled to Florence and Sienna where he seems to have stayed for over a year, learning Italian and practising and perfecting his skills as a fencer and swordsman, skills he would use to his advantage on at least two occasions in the future. In 1621 a book was published in Sienna on the subject of arms, and divided into three sections: pike, halberd and musket. The book is dedicated to Digby, and includes an engraving of him, and in it he is praised for his skills with the different weapons – praise indeed for an eighteen-year-old. Whilst in Florence, Digby also met Galileo, the famous astronomer and mathematician, who at the time was attempting to persuade the Catholic church that the earth actually revolved around the sun. It was during his time in Florence that Digby claimed to have met a Carmelite Friar who had brought 'from the East', the recipe for what became Digby's 'Powder of Sympathy' (of which more later). In 1623 Digby travelled to Madrid to meet his cousin, Sir John Digby, who was still England's ambassador to Spain, and where he met and befriended Charles, the Prince of Wales, who was there with George Villiers, 1st Duke of Buckingham (the King's favourite and quite possibly, lover), attempting to win the hand of the Infanta, Maria Anna, daughter of the late Philip III of Spain (who had died in 1621). Because Digby was a Catholic, he was asked to negotiate with the Bishop of Toledo, Primate of Spain, but despite the Infanta taking a liking to Digby, her dislike of both Charles and Buckingham was

---

[7] In 1634 Digby donated Allen's books to the Bodleian Library along with a substantial number of his own.

such that the negotiations did not go well. This was partly due to Charles, having heard that the Infanta walked barefoot in a private garden every morning and, desperate to meet her in person, hid up a tree whose branches hung over the garden wall so that he could jump down into the garden as the Infanta passed by. This he did, rolling through the rose bushes as he fell, accosting the Infanta and frightening her companions so that they all ran away. Not a great start to the relationship.

Whilst Digby was in Madrid, he was returning to his lodgings one evening and he, his cousin and a friend were attacked by a band of local ruffians (said to be fifteen strong). In the ensuing sword fight, Digby killed two of the attackers, winning him both praise and notoriety in equal measure. The attempt by the Prince of Wales to marry the Infanta finally fell apart when Philip IV (Maria Anna's brother) demanded that Charles convert to Catholicism (and remain in Spain for a year after the wedding). Charles and his retinue were so outraged that they set sail for England and upon arrival demanded that King James immediately declare war on Spain.[8] Whilst in Madrid, Digby also met James Howell, who had been an employee and protégée of Sir Robert Mansell at his Broad Street glassworks. It would be fascinating to know if Digby and Howell discussed glassmaking. Digby and Howell are discussed in more detail at the end of this chapter.

Charles then turned his attentions to France and married Henrietta Maria, the daughter of the King of France, Henry IV, whom he had met in 1623 (when she was 14) on his way out to Madrid. The happy couple were married by proxy in front of the doors of Notre Dame in Paris in May 1625, with Charles *in absentia*, soon after he ascended to the throne. They didn't actually meet as man and wife until June 1625 in Canterbury, and it is said that Charles delayed calling his first Parliament until the marriage was consummated in case Parliament objected to it, as the fifteen-and-a-half-year-old Henrietta Maria was, of course, a Catholic.[9]

Digby returned to England with the Prince of Wales, arriving at Portsmouth on 5 or 6 October 1623. After a short illness and a visit to his mother, Digby presented himself to James I at Hinchingbrooke House near Huntingdon, where he was knighted on 23 October 1623 (some reports say the 16th, the 21st, or the 28th), and made a Gentleman of the Bedchamber to the Prince of Wales. In 1624 he received a degree from Cambridge University upon the occasion of a visit by the King to the

---

[8] It is interesting to note that Sir Robert Mansell was one of the twelve people tasked with looking into the breakdown of the proposed marriage of the Prince of Wales to the Infanta, the 'Spanish Match', as it was known.

[9] Henriette Maria was the wife of one King, Charles I, the mother of two Kings, James II and Charles II, and the grandmother of two Queens, Queen Mary (of William and Mary fame), and Queen Anne.

The Knight Who Invented Champagne

> Introduction to *The Private Memoirs of Sir Kenelm Digby*, 1827.
>
> 'On the accession of Charles the First, Sir Kenelm Digby became one of the chief ornaments of Whitehall. Charles, who did not love gaiety, highly esteemed him, however, for his admirable talents; but to the Queen, who before her misfortunes had a very lively disposition, he rendered himself infinitely agreeable, and she seems to have conceived a friendship for him which lasted through life. He was a party in all the royal diversions, which indeed he frequently planned and directed; and such were the volubility of his spirits, and the careless elegance of his manners, that it should have seemed that he had been bred from his infancy in a court.'

University (despite not having studied there). On 27 March 1625, King James died and the Prince of Wales ascended to the throne, making Digby a member of his first Privy Council. In order to do this Digby officially converted to Protestantism, as many others had to do in order to keep themselves in favour with the monarchy. Thus, knighted and honoured, he became an important and valued member of the new king's retinue and applied his mind to finding himself a wife.

Digby gets married

In 1615, when Digby was only twelve, he had met Venetia Anastasia[10] Stanley (1600–1663), a Catholic, the grand-daughter of the Earl of Northumberland, and then aged fifteen. Her mother, Lady Lucy Percy, had died when Stanley was only a few months old and her father had entrusted her upbringing to a family who lived at Enstone Abbey[11] in Oxfordshire and who had been great friends of her mother's. She was evidently an extremely pretty young woman, said to have the smallest waist in London, and Digby seems to have taken a shine to her – one report says he 'philandered about her for a year or two' – and her memory stayed with him. His mother, however, was against the match and it was one of the reasons that he had been sent on the Grand Tour. The news of his 'death' had travelled to England, where Stanley heard about it and believed it to be true. Digby, who still harboured fond memories of Stanley,

---

[10] There is some doubt about whether Anastasia was actually her middle name.
[11] The church in the village of Enstone is (coincidentally) named after St Kenelm, a saint from the 900s.

CHAPTER 7

had been writing to her whilst he was away, but the letters were intercepted by his mother as she was determined to end the relationship. Stanley had moved to London before her twentieth birthday, where she gained a reputation for being rather free with her favours. The noted diarist of the time, John Aubrey,[12] said that she was known for her *bona robe* (curvaceous figure), her 'licentious looks' and that she had been the 'concubine' of Richard Sackville, the Earl of Dorset by whom she bore at least one, if not two, of his several children.[13] Their relationship had been such that he settled a lifetime income on her of £500 a year, a considerable sum in those days. When Digby reappeared and reacquainted himself with Stanley, the Earl stopped paying her the income, whereupon Digby sued the Earl, winning the case and reinstating the payments. However, the Earl died in 1624 and shortly afterwards, in January 1625 – the exact date is unknown as the marriage was conducted in secret – she accepted a proposal of marriage from Digby and they were married.

To say that Digby was enamoured of her is probably understating the case. Maybe besotted would be a better word. Throughout their marriage, which ended in her untimely death in 1633, but during which she bore him four sons, he kept copious notes and jottings about her which were later published as part of his *Private Memoirs*. Aubrey states that Digby married her 'much against the good will of his mother' but saying that 'a wise man, and lusty, could make an honest woman out of a Brothell-house.' Stanley bore him a son, Kenelm, fairly soon after they were married and a second son, John, on 29 December 1627. As Digby had already left on a voyage (see below) when his new child was born, he asked Stanley to make their marriage public upon the birth. His wife gave birth to two more sons, George, and Everard, the last of whom died in infancy. Stanley seems to have been happy to have married Digby and it is said that upon the occasion of the King's marriage to Henrietta Maria in 1625, when Digby and Buckingham travelled to Paris, she pawned some of her jewels so that he could afford to buy some 'finery' for the trip. Why he could not afford 'finery' out of his own substantial income is an interesting question.

---

[12] John Aubrey 1626–1697 wrote, amongst other works, *Brief Lives*, a three-volume collection of biographical notes on the prominent people of the day. It was written between 1680 and 1693, deposited by Aubrey in the Ashmolean Museum (but is now in the Bodleian Library), but it was not fully published until 1813. The books contain biographies of Everard, Kenelm and Venetia Digby and, given that Aubrey knew them personally, one might expect a fair degree of accuracy. However, Aubrey was known to have spiced up some of the entries in order to make the books more saleable, so caution about their accuracy is advised.

[13] The Earl's first marriage to Lady Anne Clifford ended, as he had numerous affairs and he was known as 'one of the seventeenth century's most accomplished gamblers and wastrels'. Before they were divorced, she bore him five children, none of the three boys surviving to inherit their father's title.

The voyage to Scanderoon

In 1627 Digby embarked on one of the most amazing escapades of his already eventful life. He proposed to obtain the authority of the new king, which was to be issued under the Great Seal of England, and which would allow him to set sail as a 'privateer' (and not a 'plunderer') and embark upon a buccaneering, money-making venture. The stated purpose of the escapade was to 'ruin the Venetian trade in the Levant, to the advantage of English commerce', but would take in a bit of smashing and grabbing en route. He proposed to sail to the far eastern end of the Mediterranean, to a port then called Scanderoon (Alexandretta and, today, the Turkish port of İskenderun), where he would attack the Venetian naval fleet. Despite opposition from Buckingham, he received the King's authority on 13 December 1627 and, having raised funds from various backers, and hired two ships and a complement of sailors, set sail from Deal on 7 January 1627/8. His two ships, the *Eagle*[14] of 400 imperial tons and the barque *Elizabeth and George* of 250 imperial tons, both under the command of experienced captains, headed for the Mediterranean.[15] This book is not the place to recount the whole story of this 14-month voyage but the highlights are worth mentioning. Having struggled down the west coast of France and Spain, battered by Atlantic gales and with his crew laid low by illness, Digby managed to engage and plunder several Dutch and Flemish ships near Gibraltar. He then entered the Mediterranean and reached Algiers harbour in February 1628. Here he gained an audience with the Pasha of Algeria, Hassan Khodja, and feasted with some Dutch 'corsairs' – essentially privateers like himself – and learnt that there were around fifty English and Irish sailors who were being kept by the Pasha as slaves. He then negotiated their release for £1,650 (possibly as much as £350,000 in today's money). He left Algiers on 27 March 1628. After he returned to England, he attempted to get the £1,650 paid back by the authorities, but they were reluctant to do so. He would have to wait until 1663 before he recovered his funds. Whilst in Algiers he made sure his ships were disinfected 'with vinegar and garlic', spent time sampling the local food – including couscous and melons – and looked for ancient texts for sale, one of which is now in St John's College, Oxford with notes on the margin about how to cook bananas in syrup, plus the wise advice about 'three things

---

[14] The *Eagle* was subsequently renamed the *Arbella* (sometimes called the *Arabella*) and was John Winthrop's flagship when he sailed to Salem to found Massachusetts in 1630.
[15] Joe Moshenska, in his book *A Stain in the Blood*, says that Digby armed himself with Captain John Smith's *Seaman's Grammar and Dictionary* 'a sort of 'Privateering for Dummies'. John Smith was, of course, famously saved by Pocahontas (or maybe not – Smith is said to have made this up).

that must not be lent: a comb, a toothbrush, a slave-girl'.

After leaving Algiers, he set sail for the real object of the voyage, the destruction and/or plundering of the Venetian fleet. He arrived off the town of Scandaroon on 10 June 1628 and found the Venetian fleet in the port, together with French and Egyptian merchant ships moored in the bay, which, perhaps unwisely, fired on his ships to warn him away. Digby waited a day until 11 June, (possibly) his birthday, to retaliate and then open fire. According to contemporary reports, the fighting was fierce and there were losses on both sides. However, Digby managed to get away with his ships 'loaded to the gunwales' with plunder, and set sail for home. On the way back he wrote an account of the 'brave and resolute Sea-Fight' which he must have sent overland to London, as the news of what had happened at Scandaroon had already reached Woolwich when he arrived back.

On the way back from Scandaroon he stopped at the island of Delos, where he took on board a large number of 'marbles' and other sculptures, some being so large as to warrant the setting up of a hastily-devised crane made out of the 'mastes of ships' to get them on board. These antiquities were eventually displayed in London and some were given (some reports say 'sold') to the King. He was then forced to call in to the small Greek island of Milos (about halfway between Athens and Crete) and lay up for a while whilst repairs were carried out. During his time on the island, he started to write his *Private Memoirs* and used this as an excuse to steer clear of the local women, who expected visiting mariners to court them, woo them and undoubtedly pay them. Digby was criticised for ignoring the locals but it is said he wanted to keep faithful to the wife he had only just married. Digby arrived back at Woolwich on 2 February 1628/9.

Digby after the voyage

On Digby's return from his privateering, having achieved a position of both notoriety and respect, he set about gaining positions of influence and power. He was made a Governor of Trinity House in ca 1629[16] (the organisation responsible for navigation buoys and lighthouses) and received the 'Patents of a Monopoly of Trade for Sealing Wax in Wales and Adjacent Border Counties'.[17] This monopoly was, in the days

---

[16] It is not known exactly when Digby was made a Governor of Trinity House as the records for these years were lost in the Blitz.
[17] Despite extensive searches, no records exist of this appointment. It may therefore be an unreliable part of the Digby legend.

when all official documents, contracts and agreements required sealing with ribbons affixed with (official) sealing wax, very lucrative. In October 1630, Sir John Coke, Deputy-Treasurer of the Navy Board, appointed Digby to the position of a Principal Officer of the Board, a position that required him to be a Protestant. In this position, and with his recent seafaring experiences, he was able to help the king in reforming and improving the Navy. Between 1632 and 1637, a total of 11 new ships and pinnaces were added to the Navy. Digby resigned from the Navy Board in 1633 after his wife died.

'Sir Kenelm Digby first invented the glass bottle'

It is around this time that it is said that Digby involved himself with, or was connected to, glassmaking, particularly bottle-making. It is, somewhat curiously, not a connection that he made himself, nor, as far as is known, mentioned in any of his writings, copious thought they are. But it is a claim backed up by a 'long and serious consideration and examination' conducted by the Attorney General and affirmed in the House of Lords, so it cannot be dismissed lightly as just speculation.

On 6 September 1661, John Colnett (Jean Colinet),[18] a glassmaker, received letters patent for the:

> perfection of makeing of glasse bottles and of glasse vessel for distillation called bodies heades and receavors of a peece being their owne invention and never before done or used in these our kingdomes.

This patent, which came before Parliament for ratification in early 1662, gave Colnett the sole right to make 'glasse bottles and of glasse vessel for distillation' for a period of 14 years, provided that the bottles contained 'the full measure of gallons, pottles, quarts and other measures' and that the bottles 'shall be marked with Colnett's particular stamp or mark'. In addition, if any were to be found that

---

[18] In Hartshorne's *Old English Glasses*, the name of Henry Holden is shown before that of Colnett. Holden ran the Savoy glassworks and was Colnett's glassmaking partner. He was later appointed to be 'Glassmaker to the King' and was allowed to put the royal coat of arms on the glasses he made. In the *London Gazette* of 16 April 1683, he advertised that he 'did not use any noxious ingredients in making all sorts of glass'.

contained 'less than their proper measures' and 'did not bear the particular stamp or mark' after 20 May 1662, then the vendors or users would be fined £5 and any bottles 'which shall be extant after 20 June 1662' would be 'seized by warrant from a Justice of the Peace and broken'.

The issuing of these letters patent caused consternation in the world of glass bottle-making. At this stage, four glassmakers, who said they were former employees of Digby, made statements to the Attorney General, Sir Geoffrey Palmer, that 'John Colnett, by false allegation, had obtained a patent' to make bottles. The 'glasshouse four' also stated that they had worked for Digby, making glass bottles, 'thirty years since', i.e. thirty years ago. The four glassmakers, firstly John Vinion and Robert Ward, but later joined by Edward Percival and William Sadler, said under oath that:

> Sir Kenelm Digby first invented glass bottles some thirty years since, and employed Colnett and others to make them for him, and they have since been frequently made by him [Digby] and also by the petitioners.

The quartet also said that Digby had later abandoned his interest in glassmaking and that the trade (of making bottles) 'had been of Publique use at several glasse houses in England'.[19]

As has already been said, the Attorney General conducted 'a long and serious consideration and examination' of the claim and on 2 April 1662 certified that Colnett was not the inventor of the glass bottle, and that 'the making of glass bottles is no new invention, for it has been of trade and public use for nearly thirty years'. Whether the Attorney General asked Digby himself for evidence is not known, but nobody has ever mentioned it or written about it. Digby was living in Covent Garden and was active in the nascent Royal Society, so surely would have heard that he was the subject of an investigation by the Attorney General into what he was supposed to have done or what he had 'invented' 30 years earlier? Another bit missing from the jigsaw.

What truth there is in the claim by these glassworkers, who said that they had worked for Digby, that 'he first invented glass bottles' some thirty years before 1662

---

[19] Hartshorne in *Old English Glasses* states in a footnote (page 221) that 'it is not true' that Digby invented glass bottles, and goes on to state the obvious, i.e. that bottles had been in use for many centuries before 1632, but then does discuss 'Digby's bottles' hinting that he might have developed a certain type of bottle, ones with an 'ancient bag- or purse-like shape'. The word bottle comes from the German *beutel* which means bag or scrotum. The traditional bottle shape for wine from the Franken region in Germany is called a *bocksbeutel* meaning 'ram's scrotum' after its shape.

Very early shaft-and-globe bottle

is of course open to question. However, we do know that the four men came forward and made the claim, under oath, to the Attorney General, and that he thoroughly investigated the matter. Perjury was, as it is today, a serious offence and we do know that Colnett did not take the matter any further as 'he offered nothing material in opposition'. Sir Geoffrey Palmer was a man with a long and blameless career in the law. He was a Royalist, who was Solicitor General in 1645 but was captured after the fall of Oxford during the Civil War and forced to sell his estates for £500. On the Restoration of the monarchy he was made Attorney General, knighted, and made a baronet. There is nothing to assume that he conducted other than a fair examination into the glass bottle business and came to the conclusion, one that has come down through the ages, that Sir Kenelm Digby did indeed 'invent' (maybe 'perfect' would have been a better word) the glass bottle in or about 1632.

Eleanor Godfrey, in her seminal book *The Development of English Glassmaking 1560–1640* (page 229) sets out the case for Digby. She says that 'all parties concerned' (in the Attorney General's investigation) agreed that there had been significant changes in bottle-making in around 1630–1632 and that 'Digby was the true inventor' of the (new) glass bottle. She says that the changes in bottle-making at the time are obvious and that the new Digby-style bottles 'are heavy, strong, and globular in shape with a high "kick" in the bottom and a long tapering neck ending in a collar for tying down corks. The colour was always very dark, varying from dark olive to brown or black'. She continues:

> The new type of bottle seems to have been made by a formula developed after the invention of the coal process in glassmaking. The wind tunnels in the new furnaces produced a higher internal furnace heat, and if the covers were omitted from the pots, increasing the temperature within them still

further, a batch [of glass] with a higher silica content and less potash and lime could be melted. Though the metal [the molten glass] was darkened by the coal fumes, this was not considered a disadvantage; in fact, it came to be regarded as a sign of superior strength.

On this evidence alone, therefore, we can assume with some degree of certainty that Digby had a definite involvement with the development and manufacture of glass bottles. Where exactly, and how exactly, history does not (yet) tell us. We know that glass seems to have been a recurring interest in Digby's life. He knew both Mansell and Mansell's protégée James Howell and had every opportunity to visit and work with them on glass development at Mansell's Broad Street glassworks, which was near Digby's house. There are several references to Digby's involvement with a bottle-making glassworks at Newnham on Severn and he certainly knew Sir John Winter, who owned coal mines, and ironworks and was very active in business in this part of Gloucestershire. Digby also knew Merret and owned a copy of the *Art of Glass*, published in English in 1662. In short, there would appear to be every opportunity and enough connections for us to conclude that Digby did have an interest in glass and in bottle-making. The timing of 'nearly thirty years' from April 1662 would put Digby at his glassmaking round 1630–1632 which fits in with his life. It could be that his interest in glass started in Florence when he was there in 1620 (and where the author of the *Art of Glass*, Antonio Neri, had lived, although he died in 1614), or maybe it was when he met Howell in Madrid in 1622. Without any documentary evidence, we cannot be sure.

Death of Venetia Digby

'Venetia, Lady Digby' on her deathbed by Sir Anthony van Dyck

In 1633, Digby's world suddenly came apart: his beloved Venetia died. There is considerable doubt about the actual cause of her death and there are several theories. The most commonly cited is that she died from taking what was known as 'viper wine', which she was taking to improve her complexion. This concoction appears to have been made by apothecaries using various ingredients. Eyre states that the tonic contained 'iron from the viper's blood, ascorbic acid [vitamin C] from the tartar, artificial hormones from the urine of pregnant mares, and hope, faith and comfort from the dribble of delicious opium'. Dr John Fulton, Professor at the Department of the History of Medicine at Yale University, states that her death was 'probably of phthisis', what today we would call tuberculosis. Another explanation for her death is that Digby himself prescribed the viper wine to his wife and made her drink it, presumably to keep her looking young and beautiful.

On the night of 30 April 1633, as reported by her maid, Venetia Digby went to her bed (the Digbys did not share a bed, as was very usual for married couples in those days) and everything appeared to be normal.[20] However, in the morning when her maid tried to wake her, she was discovered to have died in the night. She was only 33. Two things, both unusual, then took place. Digby very quickly sent a message to Sir Anthony van Dyck, with whom he had become friends, to attend their house (in Covent Garden) and paint her portrait. Whether van Dyck did this by making preliminary sketches, a quick 'cartoon' which he later finished in his studio, or whether he erected his easel, got out his palate and oil paints and painted her portrait there and then, history does not record. The portrait shows a relaxing Venetia, propped up in bed in her night attire of nightgown and mob-cap, with one eye almost half-open, looking as if she has either just woken up or is just about to blow out the candle and

---

[20] Some reports say Digby was 'tinkering in his laboratory until the early hours' and slept there so as not to disturb his wife.

nod off. The second unusual occurrence was that Digby had an autopsy carried out on his wife. (Some say that it was at the insistence of the king, who wanted to make sure his friend was in the clear.) Whether this was an attempt to discover why she had died, out of an interest in the process of the autopsy itself, or to have some proof that he did not murder her (as he was later accused of doing) we will never know. The autopsy was carried out by Dr Théodore de Mayerne, 'first physician' to the king and queen, who said that 'she had but little brain' and concluded that she died of a cerebral haemorrhage.[21] Digby also had casts of her hands, feet and face taken, again, quite common in those days, which it is said he kept close by him for the rest of his life. He treasured the van Dyck

Christ Church, Newgate, London

painting of his wife, saying that 'it standeth all day over and against my chaire and table … and att night when I goe into my chamber I sett it close by my beds side, and by the faint light of candle, me thinkes I see her dead indeed'. He also said that van Dyck 'hath altered or added nothing about it, excepting onely a rose lying upon the hemme of the sheete.'[22] Digby buried his wife in Christ Church, Newgate[23] where he erected 'an elaborate monument' to her.[24]

Whatever the cause of death, she was definitely dead, sending Digby into a sudden and deep depression. Shortly after his wife's death, he left his house and moved himself – seemingly lock, stock and barrel – to Gresham College in Bishopsgate. He also gave up his position on the Navy Board. Gresham College had

---

[21] The Royal College of Physicians' biography of the good doctor says that in 1654/5 he died of drinking 'bad wine' in the Strand, although it also adds that he was 'drinking in moderation'.
[22] The picture is now in the Dulwich Picture Gallery.
[23] Christ Church, Newgate (now known as Christchurch Greyfriars Church) was consumed by the Great Fire in 1666. The church was rebuilt to a Wren design in 1704, but the body of the church was destroyed in the Blitz in 1944 and only the side walls and a grand tower survive today. Digby was also buried there.
[24] Ben Jonson, the playwright and poet, at the time, second only in fame to Shakespeare, wrote and read an eulogy to Venetia at her funeral. Digby was Jonson's literary executor.

Gresham Place in the 1600s

been set up in 1597 by the will of Sir Thomas Gresham, who was a banker, founder of the Royal Exchange and 'financial agent' for four successive monarchs: Henry VIII, Edward VI, Mary I and Elizabeth I. The College was an educational establishment, the first institute for higher learning in London, and appointed seven professors in different disciplines: astronomy, geometry, physic, law, divinity, rhetoric and music. The seven professors initially lived at the College, housed in Sir Thomas Gresham's former home (the site of the NatWest Tower, which is now known as Tower 42). Despite not being formally attached to Gresham College, Digby was well acquainted with many of the professors and managed to secure permission to live there and set up a scientific laboratory and workshop where he lived and worked. He occupied 'five or six' rooms beneath the lodgings of the Divinity Reader and four of these were converted into a 'fair and large' laboratory. He employed one of the chemistry lecturers as his assistant, a Hungarian alchemist called Johannes Banfi Huniades.[25] It is said that Digby 'wore … a long mourning cloake, a high crowned hatt, his beard unshorne … as signes of sorrowe for his beloved wife'. 'He diverted himselfe with his chymistry and the professors' good conversation'. Digby lived in Gresham College for two years, 1633–1635. Whilst he was there, his steward, George Hartman, started to assemble the many recipes that Digby (and others) had written down and which would subsequently be posthumously published by his son John under the title *The Closet of the Eminently Learned Sir Kenelme Digbie Kt. Opened* of which more later.

In 1636 Digby converted (back) to Catholicism and in 1637 decided to move to France. He obtained a 'licence to travel' from the King, which allowed him to travel abroad with his two sons and three servants for three years, taking £50 with them. Once settled in Paris, he took up his pen and in 1638 wrote *A conference with a lady about choice of religion* which was published in Paris. It is said that 'he earned the reputation of being "not only an open but a busy Papist," though "an eager enemy to the Jesuits"' and was summoned to the bar of the Houses of Parliament as a

---

[25] There are about ten different ways that this person's name is spelt and I have taken the most usual one. He is also often described as 'Transylvanian', rather than 'Hungarian'. His two children, Johannes and Elisabeth are buried in St Leonards, Shoreditch. He died in 1646.

'Popish recusant' and examined until the middle of 1642. This was not before he had fought a duel in 1641 with a French nobleman, Mont le Ros, in Paris, who had taken Charles I's name in vain, and killed him. He was immediately pardoned by Louis XIII on the grounds that he was 'defending the fair name of his King' and was given a 200-man escort out of France to Flanders and thence to England. On his return he published an account of the duel under the title *Sr. Kenelme Digbyes honour maintained* which showed how 'in foure bouts hee ranne his rapier into the French Lords brest till it came out of his throate againe'.

Digby imprisoned and then into exile

Parliament did not take so well to Digby's return and ordered Charles I to banish him and, despite intercessions from his old friend, Marie de' Medici (before she died in July 1642), he was imprisoned: firstly the serjeant-at-arms shut him in Crosby Hall (then in Bishopsgate but now re-erected on Cheyne Walk, almost opposite Battersea Bridge), then in the Three Tuns near Charing Cross (just off Whitehall) 'where his conversation made the prison a place of delight' and then to Winchester House, Southwark which had been turned into a prison.[26] He remained there for two years. During his time there, as reported by Longueville (a descendent of Digby's):

> Sir Kenelm not only interested himself in his books but also in experiments in glassmaking. What opportunities and conveniences he may have had for carrying them out it is impossible to learn, but Southwark has long been famous for glass manufactures, and Sir Kenelm may have been allowed to call in some workmen from a neighbouring factory.'[27]

This period is, of course, ten years after the time when he is said to be involved with bottle making, but it does show that he was still interested in glassmaking technology.

Digby's release from imprisonment was aided by support from Anne of Austria, wife of Louis XIII and Queen of France and he was banished abroad. Digby went to Paris, where he had many friends and supporters, but banishment was something he took fairly lightly and it would appear that he travelled between France and

---

[26] Winchester House was in Clink Street, next door to the notorious 'Clink Liberty' – a much feared prison. Some of the foundations of the house can be seen next to the London Dungeon which partly re-creates the conditions of the Clink.

[27] Page 255 'The Life of Sir Kenelm Digby' by Thomas Longueville [a relative of Digby's], London 1896.

The Knight Who Invented Champagne

England fairly often, escaping notice and capture. In Paris he teamed up with the exiled Henrietta Maria, wife of Charles I and notionally still Queen of England and Scotland, who had been there since 1644. In 1645, Henrietta Maria, who had set up a royal court in exile at the Château de Saint-Germain-en-Laye, just west of Paris, appointed Digby to be her Chancellor and put him in charge of fundraising. In 1645 she made him ambassador to the Vatican and he was sent by the English Catholic Committee to Rome with a list of worthies to meet and ask for funds from. Pope Innocent X gave him 20,000 crowns to raise and equip an army to fight the Puritans but, needless to say, this did not happen. The Pope apparently took against Digby's manners and accused him of misappropriation, something which Digby vehemently denied (although the 20,000 crowns do not appear to have been returned to the Pope). He went between Paris and Rome securing funds for the Catholic cause and appears to have fallen in with some more radical Papists who did his reputation no good at all and brought him firmly into the sights of the English authorities. Returning to England once more to try and sort out his property and finances, he was soundly advised to go back to Paris and never return. With Charles I in prison and awaiting trial, he realised that things were getting serious for Catholics, especially supporters of royalty, and by the time Charles I was executed – on 30 January 1648/9 – Digby was back in Paris.

On 31 August 1649, the House of Commons voted on a question that had been asked of it and resolved:

- That Sir *Kenelme Digby* do depart this Nation, and all the Dominions thereof, within Twenty Days next ensuing; and not to return without particular Leave first had of the Parliament; upon Pain of Death, and Confiscation of his Estate real and personal.
- That the Estate and Estates of all such Person or Persons as shall or do conceal the said Sir *John Winter*, Mr. *Walter Mountague*, and Sir *Kenelme Digby*, or any of them, shall be sequestred.

Back in Paris, Digby continued to live a life dominated by his researches and writing and it was whilst he was there that he published *The Nature of Bodies* and *The Immortality of Reasonable Souls*. He visited Descartes before he died in 1650, and attended lectures given by Nicholas le Febure[28] (or le Fèvre – he styled himself several

---

[28] In 1660 le Febure was appointed to the position of Professor of Chemistry to Charles II of England, Apothecary in Ordinary to the Royal Family and Manager of the laboratory at St James's Palace. He became an English citizen in 1662 and was elected to the Royal Society in 1663.

different ways) in the company of John Evelyn who was visiting him, in 1651. He also joined a group of men of learning put together by Père Marin Mersenne, a Catholic priest and polymath, who became best known as a mathematician. The group, known as the *Académie Parisienne*, was the basis for the *Académie des Sciences* which was founded in 1666.

Digby returns from exile

In 1653 Digby was given permission to return to England on condition that he did not involve himself with Royalist plots, and in 1654 he returned. Digby had been in contact with Cromwell, and had undertaken some diplomatic tasks for him, visiting Germany (Frankfurt), Denmark and Norway and had formed what was evidently a friendship. It is said that they often dined together. For a Catholic and a Royalist, Digby's friendship and support of Cromwell seems strange, but he explained it himself by saying that he could achieve more for Catholics by being close to Cromwell and working with him, rather than attacking him. He also, of course, was desperate to get back his property and said 'my restitution to my country and estate, I owe wholy to my lord Protector's goodnesse and justice'. Digby was, in effect, an unofficial representative for English Catholics and in 1655 was sent by Cromwell to Rome to try and negotiate with the Pope for a softer stance towards England. This appears to have been unsuccessful. Cromwell died in 1658 and was buried in Westminster Abbey, but dug up in 1661, hanged and decapitated, with his head being put on a pole outside Westminster Hall where it stayed until 1685. There is now a larger than life-sized bronze statue of him outside Parliament.

Digby spent the next few years, until the Restoration of the monarchy, travelling around Europe. In 1656 he was in Toulouse, and in 1658, already suffering from the illness that would kill him, he travelled to Montpellier to spend time with 'men of learning', but also to seek a cure. He was suffering from 'the stones' – bladder stones – which were a fairly common complaint at that time. The French were experts in their surgical removal, although like all operations at that time, there were huge dangers involved. Of his better-known acquaintances, Francis Bacon, Oliver Cromwell, Louis XIV, Isaac Newton and Samuel Pepys were all sufferers. Whilst in Montpelier he read his paper 'A Discourse on the Powder of Sympathy' in the company of 'eminent persons'. This paper, which was published originally in French, but was subsequently translated into English, German, Latin and Italian, was widely read and seemingly believed. In 1659 Digby returned to live in his house at 43 King Street, London, the last house nearest to the Covent Garden market square, where he busied himself with his books, his writings and his experiments and where he established a laboratory. He also set up the first French-style salon where men of letters, science and learning could gather together to discuss ideas. He mixed with the most famous scientists and learned men of the day, people such as William Harvey, Ben Jonson, John Wallis, Robert Hooke and Robert Boyle, and together they argued for the creation of an official scientific body to promote

---

The Royal Society of London for improving Natural Knowledge

The Royal Society's foundation started with a group of scientists and natural philosophers centred around Sir Robert Boyle who first started meeting in the mid-1640s at Gresham College (and known as 'Greshamites'). Christopher Wren was the Gresham College Professor of Astronomy (appointed in 1657) and on 28 November 1660, immediately following one of his lectures, the twelve people present, known as the 'Founder Fellows' held what was the first ever meeting of the 'Learned Society'. They agreed to meet 'on wensday at 3 of the clock in the Terme time, at Mr Rooke's chambers in the college. In the Vacation at Mr Balls Chamber in the Temple.' They agreed to pay a 10/- joining fee and 'one shilling weekeley, whether present or absent'. The King approved the rules on 5 December 1660. The Society became the Royal Society and received its first Royal Charter from Charles II in 1662 (and a second one in 1663). Digby was one of the 41 members of the Royal Society who together drew up its constitution and was elected to be a member of its first Council. The Society operated out of Gresham College until 1710 when they moved to their own premises in Crane Street, near Fetter Lane. Sir Kenelm Digby and Dr Christopher Merret, both of whom were Greshamites, became 'Original Fellows' of the Royal Society.

learning. On 23 January 1660/61, Digby read his paper 'A Discourse Concerning the Vegetation of Plants' to members of the Society for Promoting Philosophical Knowledge by Experiments. This paper set out how plants grew and gave a clear description of the processes of germination, development and reproduction, all based upon his personal observation and experiments. It gave an analysis of the principles of growth and Digby noted that there was 'in the air a hidden food for life', thus for the first time hinting that plants breathed and required oxygen to survive and flourish. It is said that he carried out the experiments shortly after he moved into Gresham College after his wife's death. The paper was published in book form in 1661 in English and then in 1663 in Latin, French and German. Digby was an 'Original Fellow' of the Royal Society when it was formed in 1661/2.

Charles II was restored to the throne in 1660 and on 25 May he arrived at Dover, reaching London on his 30th birthday on 29 May. However, Digby's relationship with Charles II gradually deteriorated and, although he was still Henrietta Maria's Chancellor and visited her often at Somerset House (it is said he still made possets for her), by 1664 things had reached a turning point and he was banished from the court. Before this happened, he did manage to get repayment of the ransom he had paid in Algiers, but the more the king learnt of his relationship with Cromwell, the less he trusted him.

Death of Digby

In 1665, with his 'stone' troubling him, Digby set off for Paris where there were 'good doctors for the disease' and had himself borne on a litter down to Dover. However, realising that he would not last the journey, he turned round and had himself taken back to London. He died at home on 11 June 1665, the (possible) date of his birth and the anniversary of his victory at Scandaroon. He was buried, with his beloved Venetia, in Christ Church, Newgate. Richard Farrar wrote this epitaph:

> Under this tomb the matchless Digby lies,
> Digby the great, the valiant, and the wise;
> This age's wonder for his noble parts,
> Skill'd in six tongues, and learned in all the arts.
> Born on the day he died, the eleventh of June,
> And that day bravely fought at Scandaroon.
> It's rare that one and the same day should be
> His day of birth, of death, of victory!

After Digby's death, his heir, John, who had fallen out with his father, inherited precious little of value, save for Digby's copious writings and paperwork. Digby had always been an avid book collector and his vast library was still in Paris when he died. On his death, the library was claimed by Louis XIV who sold it for 10,000 crowns to Kenelm Digby's cousin, George Digby, the 2nd Earl of Bristol, who then amalgamated it with his own library and sold it in London in 1680. The sale was of 3,685 lots and made £904 4s. 10d.

The Closet of the Eminently Learned Sir Kenelme Digbie Kt. Opened

John Digby's inheritance, apart from the baronetcy, included his father's vast collection of recipes for both food and drinks which he caused to be published in 1669 called *The Closet of the Eminently Learned Sir Kenelme Digbie Kt. Opened*[29] (more commonly known as *The Closet Opened*). This ragtag of recipes, many apparently supplied by Henrietta Maria, is an amazing treasure trove of domesticity and opens up the world of mid-seventeenth-century eating and drinking. The work of putting the recipes together was entrusted to Hartman (Digby's steward), who apparently added a few of his own and undoubtedly edited and corrected those that had been assembled over the years. Apart from many other recipes and eating and drinking tips, *The Closet Opened* is the first place which recommends eggs and bacon for breakfast: 'two poched eggs, with a few fine dry-fryed collops of bacon are not bad for breakfast, or to begin a meal'. The book mentions the word 'bottle' 98 times: strong bottles, quart-bottles, great double-glass bottles that hold a quart, pottle-bottles[30], glass bottles stopped with ground stoppels of glass, stone bottles. In fact, almost every page has a reference to bottles and bottling, although nowhere does Digby say anything like 'which I helped make' or any clues to his involvement (if any) with bottle making. The book also shows Digby's interests in distilling various liquids and making cordials and *aqua composita* (tonics, perfumes, *eaux de toilettes* and other distilled preparations) all of which required bottles. There are, for instance,

---

[29] The complete title of the book is: *The closet of the eminently learned Sir Kenelm Digbie Kt. opened: Whereby is discovered several ways for making of metheglin, sider, cherry-wine &c. together with excellent directions for cookery: as also for preserving, conserving, candying, &c.*

[30] A 'pottle' is a measure of volume of approximately half an Imperial gallon. It can also mean a small container for holding fruit or other foods.

about 45 recipes for mead of one type or another, and about the same again for making metheglin.[31]

Digby's donation to the Bodleian and Harvard Libraries

In 1634 Digby donated books to the Bodleian Library, which constructed a new west wing to house them. This donation included two hundred books given to him by Sir Thomas Allen. The donation was introduced by Archbishop Laud and Digby hoped it would establish his name in learned circles. Digby also gave manuscripts to St John's College at the start of the Civil War. In 1655 Digby donated forty volumes to the Harvard College Library, established in 1638, as he wanted to entice his friend John Winthrop Jr, first governor of the Massachusetts Bay Colony, back to London to join the 'Universal Laboratory' Digby planned to set up

James Howell (ca 1594-1666)

James Howell was born and brought up in Wales: his father was the rector at Abernant in Carmarthenshire. Howell went to Jesus College, Oxford, graduating BA in December 1613. Whilst at Oxford his tutor was Sir Robert Mansell's nephew, Dr Francis Mansell, who later became Principal of Jesus College at various times. When Sir Robert Mansell set up his glassworks at Broad Street in London in 1616, he employed Howell as his 'steward' (manager). By 1618 Howell had a warrant enabling him to travel abroad[32] on behalf of Mansell for three years for the purpose of researching other glassmaking regions and factories, searching for supplies of materials and seeking out foreign experts who might want to come and work in England. Letters from Howell show that he was in Holland in 1618, and in Venice in 1619 where he visited the island of Murano and saw glass being made and where, according to Moshenska, 'he had been trying to learn the secrets of glassmaking'. He also went to Spain for Mansell to buy *barilla*, the soda-rich saltwort used as a flux in the best-quality glass.

---

[31] Mead is an alcoholic drink made from fermented honey, and metheglin is mead flavoured with herbs and spices.
[32] His permission allowed him to travel everywhere, 'but not to Rome or St Omer', possibly because of their Catholic connections.

In 1621 Howell wrote to his old tutor, Dr Francis Mansell, saying that he would be 'quitting the glass business' which he says 'will be too brittle a foundation for me to build a fortune upon'. He said that Sir Robert Mansell 'hath melted vast sums of money in the glass business which is a business more suited to a merchant than a courtier'. Upon his return from Europe, Howell stopped working for Mansell. Some reports say that Howell fell out with Mansell, but there is no evidence to support this. Howell looked at various posts which might use his newly learnt language skills, including working at the embassy in Constantinople and as a tutor, but nothing came of either.

Late in 1622, Howell was sent on a special mission to Madrid to 'obtain redress' on behalf of the Turkey Company, owners of a 'richly laden' ship called the *Vineyard* that had been seized by the viceroy of Sardinia. Sir Charles Cornwallis and Lord John Digby, 1st Earl of Bristol, who was England's ambassador to Spain (as well as being Sir Kenelm Digby's cousin) had already tried to get this dispute settled, but gave Howell a chance to pursue their claim. This he did and he appears to have persuaded the Spanish to find in favour of the English, although this didn't result in any financial compensation, as the viceroy 'proved insolvent'. Whilst in Madrid, he met the royal party negotiating the possible marriage of Charles, Prince of Wales, to the Infanta of Spain, but owing to complications (the Infanta did not like Charles and Charles did not like the Spanish king's terms) Howell's mission had to be aborted. However, his time was not wasted as, according to the 1911 edition of the Encyclopaedia Britannica, whilst in Spain, 'Howell made many friends amongst the Prince's retinue' which, of course, included Digby. Maybe in the margins of the marriage talks Digby and Howell found time to discuss glassmaking; who knows? Whilst in Madrid, Howell sustained an injury to his hand whilst attempting to break up a duel, and asked Digby if he could attempt to cure his wound using the Powder of Sympathy (see Appendix IV).

Whether because of the Powder of Sympathy or despite it, the wound healed and Digby and Howell cemented their friendship. It would be fascinating to know how far their discussions on glassmaking went. There is no doubt that Howell had considerable practical knowledge of the subject and had conversed knowledgably with many different glassmakers. We also know that Digby knew his uncle Sir Robert Mansell who had been seriously engaged in glassmaking since joining Sir Edward Zouche's partnership in 1615. Maybe Digby did get some ideas from Howell about how to improve glassmaking and maybe he did go and see his uncle and get involved with one of his glassworks? Without some documentary evidence that Digby and Howell discussed such matters, one will never know. Howell returned to England in 1624.

Chapter 7

Howell was secretary to various members of the nobility, and to the Member of Parliament for Richmond in 1627. He became secretary at the English embassy in Denmark in 1632 and was appointed by the king to be a clerk to the Privy Council in 1642. By 1644, with the Civil War having started, Howell was sent to the Fleet Prison for 'royalist leanings' (some said it was for debt) and he remained there until 1651. After he was released, he appears to have earnt a living by writing, at which he was, by all accounts, quite successful. In 1661, with Charles II on the throne, he was appointed to the post of 'Historiographer Royal of England', a position specially created for him at a salary of £100 a year. He died in 1666.

Chapter 8

Newnham on Severn

The Gloucestershire village of Newnham on Severn is reputed to have been one of the first places in Britain to use coal for firing a glass furnace, although Mabel Woods in her book about the village disputes this.[1] It is also said to be the place where Sir Kenelm Digby was involved with making bottles in a glassworks owned by, or operating under a licence from, Sir Robert Mansell. Whether these two claims are correct goes to the heart of the Mansell-Digby-bottle question.

Herbert Penn, in his *Glassmaking in Gloucestershire*, summarises the evidence for glassmaking in Newnham at this time. He states that Mansell's involvement in glassmaking is contained in Dr Nicholas Herbert's transcription and translation of the record of the Proceedings of the 1634 Forest of Dean's Eyre.[2] This near-contemporary record states that: 'Rowland Ferrice, late of Newnham, glassmaker, deceased, on 10th July 1613, built a glasshouse at Newnham within the Forest perambulation, and Sir Robert Mansell took the issues and profits to his own use and still takes them.'[3] This important nugget of information is also contained in Jim Chapman's article 'The Cider Industry and the Glass Bottle'. Penn quotes Samuel Rudder's 1779 book, *A New History of Gloucestershire*, which says that 'Mansell in the reign of Charles I, erected here [in Newnham] the first glasshouse in England which

---

[1] The claim was made in 1813 by a Mr William Fuller, who, having visited the village, wrote a long and descriptive account of its attractions.
[2] A forest 'eyre' was a local magistrate's court that governed all aspects of the forest, especially to do with who could graze what and where, and who was entitled to the products of the forest, both above and below ground. It had not been in use for many years, but in 1634 Charles I revived it in order to raise funds for himself.
[3] The entry in The Victoria County History – *A History of the County of Gloucester* (VCH) vol. 10, p. 42 – dismisses this claim, which is also mentioned in Defoe's *A Tour thro' the Whole Island of Great Britain*, but it is also recorded in 'Notes on the Diocese of Gloucester by Chancellor Richard Parsons, c.1700', Gloucestershire Record Series, BGAS, vol. 19.

was worked with stone-coal, the foundations of which still remains'. (Charles I reigned from 1625-1649). Morgan and Smith in their *Newnham: Economic History* (1972), state that glassmaking was certainly established by 1662, as James de Hugh is recorded as 'glassmaker' when he got married. John Houghton (1645–1705), like Digby, a Fellow of the Royal Society, although not a contemporary, writing in his 1681 *A Collection of Letters for the Improvement of Husbandry & Trade*, states that there were two 'bottle factories' in the town. *The London Gazette* of 18 March 1706 has an advertisement in it offering a 'good glasshouse ... very convenient for making Broad Glass or Bottles is to be let or sold'. G. H. Kenyon in his 1967 book *Glass Industry of the Weald* also suggests that Mansell had a coal-fired furnace at Newnham on Severn, but he was almost certainly repeating what others had written. However, the numerous accounts referred to above would suggest fairly strongly that glass bottles were made in Newnham on Severn during the period in question and that Mansell was involved. It is also known that Mansell had coal mining interests in nearby Swansea, just a short collier-ride away.

The reasons for selecting Newnham on Severn as a place for a glassworks were twofold: access to the river and access to a source of the 'new' fuel – coal. Newnham on Severn is on the northern bank of the River Severn, around 15.5 km (10 miles) (as the crow flies) south-west of Gloucester. It is one of the first places up the Severn Estuary from the sea where the river narrows to round 400 m (0.25 mile), and in years past there was both a ford, which could be crossed at low tide, and a ferry for when the tide was too high. There had been a river crossing at Newnham on Severn since at least Roman times, and probably before, and it was on one of the old 'drover's roads' where cattlemen, driving their cattle to London, could ford their herds with safety. It was also one of the first places that anyone coming from South Wales, heading east, could cross the Severn, apart from the hazardous ferry crossing at Aust.[4] The first bridge across the Severn at the time was further upstream at Gloucester. The village of Newnham on Severn is situated high up on cliffs, well above the flood level, making it a safe place to live and work and establish a business. Despite attempts to both bridge over and tunnel under the river at this point (for the railways) it never happened and the ferry was still working until after the Second World War. Until 1827, when the Gloucester and Sharpness Canal was completed, Newnham on Severn was also the port for Gloucester, and had a shipbuilding industry, building vessels of up to 600 imperial tons.

---

[4] Aust, where the river is fully a mile wide, had a passenger and car ferry until 1966 when the first modern Severn Bridge was opened. The bridge spans the river more-or-less exactly where the ferry ran.

CHAPTER 8

Newnham on Severn was also situated close to the coal mines of the Forest of Dean which, since the 1610 Slingsby patent on the production of glass using coal and the 1615 banning of timber for glass production, had become more important as a source of fuel. The old woodland-based glasshouses started to close, moving their operations to sites where there was coal nearby, or where there was access to coal-carrying ships from further afield. Newnham on Severn ticked both boxes. The village had long been a port for ships to both import goods and to carry the products of the region away to their final destinations: timber, bark (for tanning), hides, coal, glass and wine are recorded in 1414. Much of this trade might well have been only going to, or coming from as far as, Bristol, around 50 km (31 miles) by boat. Bristol in 1600 had 20,000 inhabitants, second only to London (with 250,000) and was the most important port on the west side of the country. Bristol had been a 'county corporate' since 1373, when Edward III gave it this status, allowing it to govern itself. Ireland was also a destination for goods, especially for bark, used by the tanning industry there.

Glassmaking in the area had been known for many decades at this point. There had been a glasshouse just south of Newent (in a hamlet now called Glasshouse – what else?) some 14 km (9 miles) north-west of Gloucester, which had been in existence since at least 1599 and was built on land owned by Sir John Winter (an important link in the Mansell-Digby-bottle chain – see below for more detail) by glassmakers travelling west from their traditional Wealden areas (Kent, Surrey and East and West Sussex) where timber was getting scarcer, owing to both glassmaking and iron smelting. It is known that Henry Bridgeman, who worked for Verzelini at the glasshouse in Crutched Friars in the late 1590s, travelled to Newent and was there by 1598 when his daughter was baptized. Other Huguenot glassmakers followed, the Tysack and Liscourt families, for instance, and they all worked at the Newent glasshouse. In 1608 a coal mine was sunk at nearby Boulsdon which could have supplied fuel for a glass furnace. In 1615, when the ban on the use of wood for glassmaking came into force, some of the glassmakers appear to have travelled south to Newnham on Severn and re-established their glassmaking, also on Winter's land, using coal as their heat source.

One of the glassmakers from Newent, Abraham Liscourt, who also worked at Newnham on Severn, eventually ended up working for Mansell at his Newcastle-upon-Tyne glassworks. After the Civil War, the glasshouses at Newnham on Severn were destroyed and do not appear to have been rebuilt until the 1670s. In nearby Bristol, which was, as has already been stated, the second largest city in Britain at the time and a major port of importation of wine (in barrel, of course), bottle-making became an important industry and by the end of the 1600s it had, after London, the largest number of bottle-making glasshouses. By 1671, a glasshouse was again operating in Newnham on Severn where, according to Thomas Baskerville in the *Victoria County History of Gloucestershire*, 'they make a great store of glass'.

Sir John Winter (ca 1600-ca 1673)

Sir John Winter (Wintour, Wyntour, Wynter), who was born in about 1600, is one of the most likely connections between Mansell, Digby and the Newnham on Severn glassworks. He was the grandson of Admiral Sir William Winter who had been one of Elizabeth I's admirals and had fought against the Spanish Armada in 1588. William Winter had been granted the manor of Lydney, a village some 8 km (5 miles) westwards down the river from Newnham on Severn towards the sea. The Winters owned substantial tracts of the Forest of Dean and had invested in iron, glass, coal, timber and bark production. In 1628 John Winter obtained from the king a 21-year lease to extract 40,000 cords[5] of wood a year from the forest to supply fuel for his furnaces. The Winters had an ironworks on the riverside at Lydney Pill, a few miles downriver from Newnham on Severn, which was probably there in 1606 and by 1632–1633 had a double forge and was producing 600 imperial tons of iron bars a year. However, in 1634 the Forest Eyre found that Winter had exceeded his authority, cutting too much timber and making too much charcoal, and he was fined over £20,000 (a massive sum in those days) and his lease was terminated.

Charles I was a great supporter of the Winters, as he was desperate for funds and borrowed heavily from the family. In 1640 Charles leased most of the Forest of Dean to Winter, so desperate for money was he, and Winter agreed to pay £106,000 over six years, plus an annual fee of £1,951 for the rights over 18,000 acres of royal forest. This gave him the right to take timber and mine coal and he was said to be the second largest ironmaster in England, with 15 furnaces and 20 forges.[6] However, with the recalling of Parliament (the Long Parliament) in 1640 and the start of the Civil War two years later (1642), the Winters knew they were on the losing side, and in 1645 Winter accompanied Queen Henrietta Maria into exile, although he did

---

[5] A cord of wood is a stack of cut logs measuring 4 feet x 4 feet x 8 feet (1.2 m x 1.2 m x 2.4 m), which is 128 cubic feet (3.63 cubic metres).
[6] The agreement also gave Winter rights over many other aspects of the forest including making parishes, creating and endowing churches, and powers to execute the Forest Laws and levy fines and punishments. These rights and powers did not endear him to the forest-dwellers.

return to fight for the King. In 1648/9 he was accused of conspiring to help a Catholic rebellion in Ireland and was sentenced (along with fellow Catholics Sir Kenelm Digby and Sir Walter Montagu, who were also implicated in the plot) to perpetual exile. Refusing to leave Britain, he was imprisoned in the Tower of London for four years. After the Restoration in 1660, his manor and lands were restored to him and he resumed his activities as an ironmaster. His last years were spent trying to perfect the production of coke for smelting iron, something that was not achieved until 1709, when Abraham Darby used a coke-fired blast furnace at Coalbrookdale, in Shropshire.

Sir John Winter was not only a landowner and industrialist but also highly regarded by the King and Queen Henrietta Maria. In 1632 Winter was a signatory to the setting up of the Company of the General Fishing of Great Britain and Ireland, which was an attempt by the King to regulate and derive income from the fishing around the British Isles. Winter also served on its council. In May 1638 he was appointed to be Secretary to the Queen, a position he retained until 1642. (He was again Secretary to the Queen from 1660, the year of the Restoration of the monarchy, until 1669, the year she died.) This was the same queen, who, having gone into exile to France in 1645, appointed Digby a year later to be her Chancellor in charge of fundraising. Samuel Pepys thought very highly of Winter and had many meetings with him to do with the supply of items for the Navy: timber from the Forest of Dean for ships, and iron for cannonballs, chains and anchors from Winter's ironworks.

Given that Digby was Queen Henrietta Maria's Chancellor and Winter her Secretary, it seems inconceivable that they did not know each other well, as at times they must have worked closely together on the queen's state matters and finances. They were both at heart Catholics, one of the reasons that the Queen wanted them appointed to high positions in her household. Winter and Digby certainly had common interests in several other areas: furnaces, coal and glass production being three of them. Whether they knew each other in the period around 1630–1634 when Digby was supposedly involved with making glass bottles at the Newnham on Severn glassworks, situated on land owned by Winter, is open to conjecture, but it would seem a distinct possibility.

Chapter 9

Dr Christopher Merret 1614-1695

Christopher Merret (also often spelt Merrett) was born on 16 February 1614 in Winchcombe, Gloucestershire, 13 km (8 miles) north-east of Cheltenham. He appears to have been educated locally, and went to Oxford in 1631 to Gloucester Hall (which later became Worcester College), then moved to Oriel College from which he graduated in 1634. He received his B.Med in 1636 and D.Med on 31 January 1642/3 both from Gloucester Hall. Various sources say he was married by 1640 to Ann Jenour of Kemspford, a village about 48 km (30 miles) south of Winchcombe, but little is known about her. They had a son, also called Christopher. The family moved to London, and Merret set up a practice, was admitted as a candidate of the Royal College of Physicians (RCP) in 1648, and became a Fellow in 1651. He was Goulstonian Lecturer[1] in 1654 and a College Censor[2] seven times between 1657 and 1670. He was appointed by the College as the first Harveian Librarian (named after Dr William Harvey[3] with whom Merret was friendly, who had donated his personal library to the College and who had nominated the very well-qualified Merret for the job) and Keeper of the College for which he received a salary of 20 shillings a year, free board and lodging and freedom from College 'taxations'. In 1653–1654 the growing library was moved to a fine building in Amen Corner, just behind the Old Bailey in London, which Merret leased from the Dean and Chapter of St Paul's for 21 years, at £1 a year. Merret greatly

---

[1] The Goulstonian Lecture is an annual event given by one of the four youngest doctors in the RCP. It started in 1639 and continues until today. To be asked to give the lecture is considered a notable honour.
[2] The Censors were appointed to police malpractice in medicine in England and to make sure that unqualified doctors did not practise.
[3] Harvey was a very famous doctor and physician of the time, was 'Physician Extraordinary' to James I, 'Physician Ordinary' to Charles I and the first person to accurately describe the circulation of the blood. It is said that he formed his theories about blood circulation whilst accompanying the king on deer hunts and studying the carcases of dying and freshly killed deer.

expanded the library over the next few years, bringing in several large donations of books and papers. In 1656 Harvey made over his estate to the College, drawing up a 'deed of gift' and laying down the specific duties and tasks of the Librarian, which included keeping all 'books, pictures, statues, presses, carpets, and other utensils … cleansed, swept, and preserved from dust and misusage'. Harvey's deed of gift also stated that the Librarian should 'from time to time be named and chosen', something which Merret later disputed, saying it didn't apply to him, claiming that the appointment had been for (his) life.

The Great Plague and the Fire of London

Merret practised medicine from the Amen Corner premises during the Great Plague, which ravaged London in 1665 and 1666, staying at his surgery and helping those afflicted by the disease as long as he could until conditions got so bad for him and his family that he was forced to flee to the country. Before he left, he put some of the College's most valuable treasures and £1,000 in cash into a locked iron chest, which he left behind. Shortly after he left, thieves broke into the building, smashed open the chest and removed the contents. As one can imagine, although he had been doing his best in difficult circumstances, the other members of the College took a dim view of this loss. Having returned from the country, as the plague had subsided, Merret and his family were forced once more to evacuate their home by the Great Fire of London in 1666. The fire started on Sunday 2 September in Pudding Lane near London Bridge and on Tuesday, with it nearing Amen Corner, Merret packed up 148 of the Society's most valuable books (out of an estimated 1,300 volumes), portraits of William Harvey and Simeon Fox,[4] and, taking them with him, left the premises. On Thursday 6 September, the last day of the fire, the building and all its contents, which included not only the College's library and valuables, but also Merret's personal papers and books, were destroyed. Without premises and without a library, the College had no need of a librarian, and Merret was dismissed from his post. The College asked Merret to surrender the lease on the building to the Dean and Chapter of St Paul's for a consideration of £27 10s. of which £2 10s. was to be given to Merret as compensation for his loss and on the agreement that he should surrender the lease. In 1669–1670 Merret was paid the money and presumably accepted the terms.

---

[4] Simeon Fox was a prominent doctor and President of the Royal College of Physicians from 1632 to 1640. He died in 1642 at Amen Corner and is buried in St Paul's Cathedral. The portrait of him that Merret rescued is said to have 'disappeared'.

Chapter 9

Over the next few years, Merret fought a bitter battle with the College, taking it to the Court of King's Bench. Merret asked to see the original deed of gift under which Harvey had given his library, so that he could transcribe it and present it to the court, but he was denied permission to see it. During the legal proceedings, Merret's son Christopher gave an account of the day they left their house. He recounted that: 'The College was burnt down between 3 and 4 o'clock on Tuesday 4th and that the day before, my father had sorted out the books and put the best ones in the yard in readiness to remove them'. The son reported that 'his father had been the last person in the college' and he gave a dramatic picture of his father walking down Warwick Lane, which was on fire on both sides, with arms full of books 'followed by the bedel [beadle]'. During the case, the College demanded that Merret return all the books and other belongings that he had removed prior to the fire, which Merret was hanging onto, one assumes as a bargaining chip. However, after many court appearances and much legal argument, his case failed and the College had triumphed. He was then 'four times (in a loyal manner) … summoned by the beadle' to appear before the President and Censors, but he didn't appear, made no apologies, so on 30 September 1681 he was deprived of his membership of the College. He appealed to the Court of King's Bench against his dismissal, but again, the court found in favour of the College. Some said his dismissal was because he hadn't paid his subscriptions (something he claimed under his appointment as librarian he didn't have to do), but Monk, in his biographical sketches of past members of the RCP, doesn't mention this. However, on 22 July 1685 the minutes of a meeting of the Council of the Royal Society record that 'that severall [sic] Members of the Society, were in great Arrears, contrary to the Statutes, [and] It was Ordered, That the following names should be left out of the list next to be printed, unless they satisfied the Treasurer in the meantime'. There then followed a list of over 50 names, some of the great and good of the day amongst them, including 'Dr Merret'. It is assumed he never 'satisfied the Treasurer' by paying his arrears and was duly struck off the membership list.

The years between 1669 and 1681 were ones of considerable upheaval in the medical world, with several factions jostling for position. Merret played his part in suggesting new methods of educating doctors, and conducting a 'pamphlet war' with the old guard. He was also at the forefront of the battle against apothecaries and 'chemical practitioners', some of whom undoubtedly knew their profession but many of whom were charlatans and no better than witch doctors who plied their trade on gullible members of the public. In 1670 Merret published *A Short View of the Frauds and Abuses Committed by Apothecaries* which sums up his views on the subject. He called them 'the veriest Knaves in England' and accused them of multiple malpractices including (but not only) over-prescribing, over-charging and generally duping their patients. It was not a booklet likely to win friends and influence people!

Apart from his medical duties and interests, Merret was intrigued by, and became an expert in, several other scientific areas. He was one of a group of natural historians, physicians and mathematicians – now known as the '1645 group', centred around Sir Robert Boyle – who first started meeting in the mid-1640s. Many of the members of this group were also involved with Gresham College and included Sir Kenelm Digby. Digby and Merret were both 'Original Fellows' of the Royal Society when it was formed in 1662 and it is therefore inconceivable that they did not know each other. Digby certainly owned a copy of the *Art of Glass* (see below) as it was in the sale catalogue when his library was sold in 1680.

Merret's interest in glass

In 1662 Merret translated from Latin *L'Arte Vetraria* (The Art of Glass) by Antonio Neri, who is usually described as a Florentine monk. The book, written in 1611 and initially published in Florence in January 1612/13, was at the time the world's most famous (and possibly only) book on glassmaking and covered all the technicalities of glassmaking as then practised. It contained, for the first time in print, the secrets of making truly clear glass, *cristallo*, which the Venetians had so jealously guarded. Not only did Merret translate the original five 'books' of original text, but added an 'Epistle to the Reader', and notes 'Of the Furnaces' and various 'observations' which almost double the length of the original. At the end is added 'An Account of the Glass Drops' (also known as Batavian Tears) brought 'out of Germany' by 'His Highness Prince Rupert' which the Royal Society had investigated. They were made by dropping molten glass into water so that it rapidly cooled and would withstand blows by a hammer, and were the precursors to toughened glass, found on the front of every one of today's smartphones. The book is dedicated to Sir Robert Boyle, Gresham College.

Tom Stevenson and Merret's paper on wine

The publication that uncovered, at least for the modern English sparkling wine industry, the significance of Merret's contribution to the sparkling wine debate was when Tom Stevenson wrote about him in the 1998 edition of *Christie's World Encyclopaedia of Champagne & Sparkling Wine* and in it published (for the first time) a slice of Merret's original lecture notes. However, this wasn't the first time Merret had been mentioned in print in relation to wine: Alexander Henderson in his *The History of Ancient and Modern Wines* (published in 1824) mentions it, although not in relation to adding sugar to wine; André Simon, in his *The History of Champagne*, published in 1962, discusses the whole topic of how the English made champagne sparkling before the French, mentioning Digby and the discovery of his bottle-making skills and the addition of sugar; Patrick Forbes in his 1967 book *Champagne: The Wine, the Land, and the People* actually mentions Merret by name (p. 129) and his part in the story of making champagne the sparkling wine we know today (this is requoted by Bonal in his *Le livre d'Or du Champagne* although the good Mr B gets his name wrong and calls him 'Doctor Morret'); and Christopher Fielden in his 1989 book *Is this the wine your ordered Sir?* has the whole story. However, it was Stevenson's writing about it in 1998, and the publication of the text that brought it to the attention of today's audience.[5]

Papers on wine read to the Royal Society

Towards the end of 1662, two members of the Royal Society, Dr Walter Charleton[6] and Merret, proposed to read papers on different aspects of wine and its treatment. These are variously listed in the Journal Books of the Royal Society as 'Some Observations Concerning the Ordering of Wines' and also sometimes 'Of The Mysterie of the Vintners'. On 19 November 1662 the pair were asked to bring their papers in at the next meeting and on 26 November 1662, Charleton 'delivered in' his 'Observations Concerning the Ordering of Wines' but it would appear that there

---

[5] Hugh Johnson in his 1989 book *The Story of Wine* also mentions Merret's Royal Society paper, but this was after Stevenson's book.
[6] Charleton was, like Merret, an Original Fellow of the Royal Society, a Fellow of the Royal College of Physicians, a Censor and the Harveian Librarian. He was also 'physician in ordinary' to both Charles I and Charles II. He was the son of the Rector of Shepton Mallet, a town known then, as now, for its cider production. His name is spelt both Charlton and Charleton in the Royal Society's Journal Books.

The Knight Who Invented Champagne

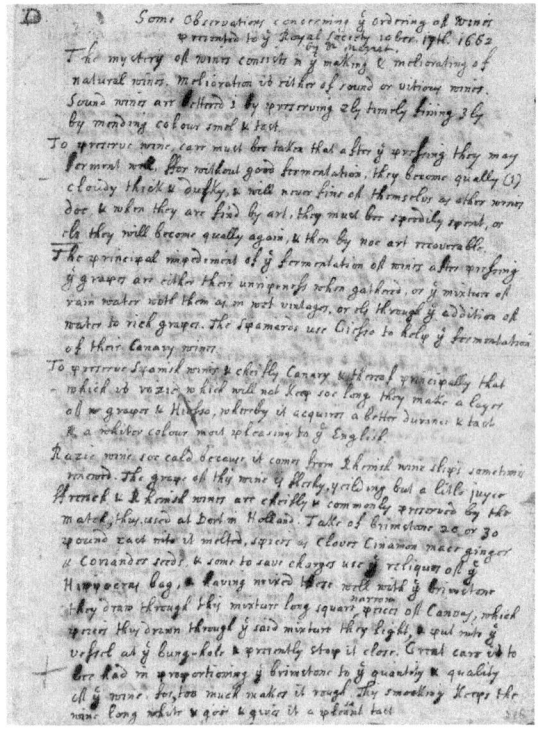

Page 1 of Merret's 'Observations concerning the ordering of wines'

was no time to read either his, nor Merret's, and they were asked to 'bring them in again' at the next meeting. On 17 December 1662, Merret finally delivered his paper to members, then 'assembled in Gresham College' (of which more below).

Despite Charleton's paper coming up for discussion on several subsequent meetings, the paper does not appear to have actually been read to members in person. On 31 December 1662 Charleton was given 'leave to keep for a fortnight' both Merret's and another member, Dr Wilde's, papers on wine, 'in order to compare them with his own, to reduce them all into one body, and return them to the Society'. On 28 January 1662/3, Charleton 'brought in again' his *The Mysterie of Vintners: or A Brief Discourse concerning the various Sicknesses of Wines, and their respective Remedies, at this Day commonly used* and the paper was 'ordered to be registered'. This is the last time these papers are mentioned in the Journal Books of the Society.

The two papers, Charleton's and Merret's, were eventually published in book form in 1669 under the title *The Mysterie of Vintners*.[7] The papers appeared sandwiched with (and after) another 'discourse' called *Concerning the Different Wits of Men* written 'At the Request of a Gentleman, Eminent in Virtue, Learning, Fortune.' The book stated that Charleton's and Merret's papers had been 'delivered to the Royal Society, Assembled in Gresham-Colledge [sic] on the 26 of November, Anno Dom. 1662' which we know was not quite true. One had been read, one had not been read, and neither on that date. Charleton's paper contains lots of tips and tricks to remedy faults and recover wines from different states, but nothing about bottling or adding sugar to bottled wine. At the end of the printed version of the paper, Charleton said:

---

[7] Printed by R.W. for William Whitwood at the sign of the Golden-Lion in Duck-Lane, near Smithfield, 1669.

Nor have I at present any thing more to add to this Essay toward a History of Wines, but my humble request to Your Lordship, and the honour'd Fellows of this Royal Society, that You would be pleased to pardon the many defects of it, and that if the Enquiries therein made come short of your expectation, You would suspend Your Curiosity untill my Copartner in this Province, the Learned Dr. Merret, shall have brought in his Observations concerning the same subject. For, I doubt not but the fulness of his Papers will supply the emptiness of mine.

Why exactly these final sentences were not edited by Charleton who knows. He certainly had enough opportunities and time. They are written as though Charleton were giving his paper in front of members and was to be followed by Merret. There is also no reference to 'Dr Wilde' whose thoughts on wine he was meant to be including.[8]

Merret's revelations concerning adding sugar to wine to make it 'brisk and sparkling'

As we know, on 17 December 1662, Merret read his paper headed (in his handwriting) *Some Observations Concerning the Ordering of Wines*.[9] The opening paragraph states:

The mystery of wines consists in the making & meliorating of natural wines. Melioration is either of sound or virtuous wines. Sound wines are bettered by preserving and by timely fining and by mending colour, smel and tast [sic].

On page 8 of his paper, the last page, Merret says:

Flat wines recovered with Spirit of wines, raysins and Sugar or Molassey and Sacks, by drawing them on fresh lees.

---

[8] There was also another, later publication called *The Art and Mystery of Vintners and Wine Coopers*, written by 'E.T. – a wine cooper of long experience who served two apprenticeships to a Vintner in the City of London'. It was published first in 1675 and in different versions over the next century. In 1735 and 1748 it was published in a version to which had been added the Charleton and Merret papers.
[9] The spellings and capitalisation in the quoted text are exactly as per Merret's handwritten notes.

Here Merret is acknowledging that raisins plus sugar of some sort makes 'flat' wines 'recover'. Flat is generally taken to mean still, i.e. not sparkling, therefore adding a source of yeasts (raisins) plus sugar would cause re-fermentation to take place. Merret then continues:

> Our wine coopers of later times use vast quantities of Sugar and Molosses to all sorts of wines to make them drink brisk & sparkling & to give them spirit as also to mend their bad tastes, all which raysins & Cute & stum perform.

To aid our understanding of this sentence, Merret has helpfully already given the following explanation for 'stum':

> *Stum* is nothing else but pure wine kept from fretting [fermenting] by often racking and matching it [matching means introducing sulphur into a barrel – see Chapter 2 for full explanation] in clean vessels and strongly scented (i.e.) new matched, by means whereof it becomes as clear or clearer than any other wine, preserving its self from both its lees by precipitation of them. But if through neglect it once fret, it becomes good wine. The bung of the vessel must be continually stopt, and the vessels strong lest they break. A little stum put to wine decaid, makes it ferment afresh, and gives life and sweetness thereto, but offends head and stomach, torments the gutts, and is apt to cause loosnesses, and some say barrenness in women.

'Stum' comes from the German (or Dutch) word *stumm* which means to mute or keep quiet, i.e. to prevent fermentation taking place.[10] The term *stumm schwefeln* is still used by German winemakers to describe the practice of heavily over-sulphuring grape juice to stop it fermenting (this juice is known as *süss-reserve*). The French for mute or silent is *muté* from which we get the word 'mutage' which is the action of fortifying wine, such as is practised in the production of port, sherry and other sweet wines. Fortification is the addition of raw alcohol at a strength of about 75 per cent (twice that of many of today's gin, vodka and whisky brands) to either completely non-fermenting grape juice – think *Beaumes de Venise* or *Moscatel de Valencia* and other *vin doux naturels* (naturally sweet wines) – or to partly-fermenting wine in order to stop the fermentation and retain some residual sweetness in the wine – as in port, sherry and Madeira.

---

[10] The English saying 'keep shtum' means to 'keep a secret' or 'keep your mouth shut' about something and is a Yiddish expression. Also sometimes spelt 'schtum' and 'shtoom'.

After Merret's lines concerning 'our wine coopers', he explains that 'cute' is 'wine boiled to the consumption of ½', i.e. wine concentrated by boiling.[11] Now this is a slightly problematic statement, as *wine* boiled to reduce it by half would result in a liquid without very much (if any) alcohol in it as it would all evaporate during the boiling process. However, *grape juice* boiled and reduced by 50 per cent would be concentrated and result in a relatively stable, very sweet liquid which could be added to a wine to both sweeten it and provide sugar for further fermentation. At the end of this section Merret adds: 'Some instead of Cute, make it of Sugar, Molossus and Honey, or mix them with the Cute' which would imply that 'it' is something sweet. From that we conclude that 'cute' is grape juice concentrate, not concentrated wine.

This revelation to the members of the Royal Society that their wine was being sweetened and made to sparkle in the glass didn't raise any eyebrows or hackles. They were all men of the world and knew that most wines (and ciders) needed 'improving', especially late in the season when wines, getting tired, had started to oxidise or, worse still, were turning into vinegar. Only the week before, the Rev. Beale had read his paper in which he specifically discussed adding raisins and sugar to *bottled* ciders to 'improve' them. Merret, of course, only mentioned 'wine coopers' and the addition of sugar to, one presumes, the vessels that 'wine coopers' dealt with, i.e. barrels. How much better would it have been for Merret to mention 'brisk and sparkling' in relation to bottled wine? It is also a pity that neither Beale nor Merret mentioned Digby in relation to bottles. Digby was back in London at this time, and whilst not in the best of health, he was on the Council of the Royal Society and is said to have been 'constant in his attendance at its meetings.' Maybe he was in the audience?[12]

Merret's other writings

Merret's best-known work (to the non-wine world), written in 1666 and published in 1667, is *Pinax Rerum Naturalium Britannicarum* in which over its 226 pages or so he lists every known *Vegetabilia, Animalia et Fossilia* in Britain at the time. Apart from some common and place names, it is written entirely in Latin. It has been criticised for being just a list of entries culled from other publications and from other people, and for five years he did employ a well-known naturalist, Thomas Willisel, to help him

---

[11] In France, *vin cuit* (cooked wine) is grape juice which has been concentrated by boiling. Definitions of *vin cuit* often refer to 'new wine' being boiled. 'New wine' refers to grape juice that has just started to ferment.
[12] At the time, the Royal Society did not keep a record of who attended their meetings.

gather information (and since when did using a researcher mean it wasn't all your own work?), However, the book did contain the first list of British birds, all 170 of them, and it became well known and respected amongst learned and scientific circles in London. The printers of the *Pinax* were in St Paul's Churchyard (the home of many printers at the time), and it is said that almost all of the first editions were lost in the Great Fire, which also razed old St Paul's to the ground. Very few copies of the 1666 edition exist and almost all are of the 1667 and subsequent reprintings. The book was dedicated to Dr Baldwin Hamey, a friend of Merret's, who was a wealthy physician of Dutch and Flemish extract who bought the lease to the house at Amen Corner and donated it to the College.

The papers that Merret presented to the Royal Society and published in their *Philosophical Transactions* are worth listing, as they show the breadth of his scholarship:
- Observations concerning the uniting of barks of trees cut, to the tree itself.
- An experiment on *Aloe Americana serrati-folia* weighed; seeming to import a circulation of the sappe [sic] in plants.
- An experiment of making cherry-trees, that have withered fruit, to bear full and good fruit; and recovering the almost withered fruit.
- A relation of the tin-mines, and working of tin in the county of Cornwall.
- The art of refining.
- A description of several kinds of granaries, as those of London, of Dantzick, and in Muscovy.
- An account of several observables in Lincolnshire, not taken notice of in Camden, or any other author.
- A table of the washes in Lincolnshire.

After leaving Amen Corner, Merret lived in a house in Hatton Garden and died on 19 August 1695, aged eighty-one. He is 'buried twelve feet deep' (some reports say fourteen feet) in St Andrew's Church, Holborn.

Chapter 10
Introduction to champagne

The idea that champagne (and for that matter all other bottle-fermented sparkling wines) were ever anything other than the product we know and love today is hard to believe, but a look back through history shows otherwise.

Vines have probably been grown in the wider Champagne region since at least Roman times (they grew vines pretty much everywhere they invaded and colonised), although, according to Stevenson, the first document to mention vines in the region dates back to AD 651. Through the ages since, the fortunes of the growers, winemakers and merchants in the region have waxed and waned. Monastic viticulture was important, as it was in so many other old-world regions, and during the twelfth to fifteenth centuries it was the monks and their *celléries*[1] who would be in charge of the decisions both in the vineyards and the wineries. Champagne became important, both as a region and for its wines, through a combination of factors. In the days before good roads, canals and railways, having navigable rivers was a great asset to a region. Champagne has access to four navigable rivers: the Aisne, the Aube, the Marne and, of course, the Seine, enabling flat-bottomed craft to navigate all the way to Paris and then, on larger ships, out into *La Manche* (the English Channel) to distribute wine to other northern European destinations. The region was also a place known for its trading links with four major Roman roads meeting at Rheims, and for the famous *Foires de Champagne* which were held in several different places throughout the year. These fairs were important to several trades, but especially for the cloth trade and particularly for selling linen sheets. They attracted visitors from far and wide who no doubt enjoyed the local wines, thus spreading their reputation. Rheims, of course, is also famous for its magnificent

---
[1] Also known as *procureur* or administrator, and responsible for running the abbey.

The Knight Who Invented Champagne

Joan of Arc

cathedral, the place where thirty-three French (or at least Frankish) kings were crowned between Louis the Pious (son of Charlemagne and co-Emperor) in 816 and Charles X in 1825. Because of its association with royalty and, of course, its relative nearness to Paris, by the middle of the 1400s the wines of Champagne were starting to be traded more widely and appreciated by people of position and authority. In 1420, the Treaty of Troyes gave the French crown to Henry V of England and his heirs (a claim the English crown did not give up until 1801), and the Earl of Salisbury became governor of Champagne. English ownership of large parts of France continued until the mid-1430s (following the trial and burning at the stake of Joan of Arc in 1431), but eventually the French reclaimed Gascony, England's last foothold in France (apart from Calais, which would remain English until 1558), following the Battle of Castillon in 1453. Constant wars and feuding did nothing to help winegrowers in the region and it wasn't until the ending of the Hundred Years War in 1453 that vineyards really became established on a commercial basis and Rheims and the surrounding villages, especially Épernay, became centres of wine production.

The relatively peaceful times after the Hundred Years War lasted until the mid-1500s when the region was once again plunged into conflict, with Catholics and Protestants fighting for religious supremacy. The Abbey of Hautvillers was destroyed in 1564 and the monks took shelter in Rheims where they stayed for forty years. However, in 1598 the Treaty of Vervins was signed between Henry IV of France and Philip II of Spain and the region could start to return to normality.

It was after this time that the wines from the Champagne region started to gain a regional identity, with the village of Aÿ[2] being referenced several times in many different

---

[2] Aÿ is a village just to the east of Épernay.

sources. François Bonal, in his book *Livre d'Or du Champagne*, states 'the term *vins de Champagne* did not appear until around 1600 and did not become commonly used until the second half of the seventeenth century'. Bonal also quotes a farming manual called the *Maison Rustique* which first appeared in 1586 which states that 'the wines of Aÿ, being less vinous than these [the wines of Gascony] are also salubrious beyond comparison.' The book continues with 'the wines of Aÿ are claret and tawny coloured, subtle, delicate and of a very agreeable flavour, and for these reasons are desired for the mouths of Kings, Princes and great Lords.' The house of Gosset, founded in Aÿ in 1584, which claims to be the oldest champagne house, was originally known for its light red wines and it wasn't until the late 1700s that it started to make sparkling wines.

There is, of course, no suggestion in any book or document of this time that the wines of Champagne were anything other than still wines. The first literary reference in French to sparkling champagne is not until the 1690s, when Madame de Sévigné (1626–1696) in one of the many letters she wrote, called it *le vin du diable* – the devil's wine – for the way it was lively in the glass. Another reference to sparkling wine from the region dates from 1700, when the Abbot of Chaulieu advises Phylis, the Duchess of Bouillon to 'Come, Phylis, spend the evening with me. See how this wine sparkles, envious of your eyes. Drown with thy hand in the sparkling foam, the worries of tomorrow'. The main reason that the wines from Champagne were not sparkling at this time is that wines of any sort were not bottled at source. Bottles for wine did not really exist, let alone bottles capable of holding the pressure of a secondary fermentation. All wines were sold in barrels as still wines, and bottling, such as there was, was carried out at the point of consumption. Most wine was sold out of the barrel in inns and taverns, or, if the household was large enough, to private houses where it was dealt with by the butler.

Dom Pérignon

No history of Champagne – the region or the wine – can be complete without reference to its most famous son, Dom Pérignon (1638/9–1715), and the part he played (as well as the part he didn't play) in the creation of the world's most famous sparkling wine. Pierre Pérignon was born in 1638 or 1639 to a middle-class family in the village of Sainte-Menehould which is about 85 km (53

Dom Pérignon

The Knight Who Invented Champagne

> According to Godinot's book, one of Dom Pérignon's secret ways of improving wines was as follows:
>
> 'A pound of sugar-candy was dissolved in a *chopine* of wine, to which was then added five or six stoned peaches, four sous' worth of powdered cinnamon, a grated nutmeg, and a *demi septier* of burnt brandy; and the whole, after being well mixed, was strained through fine linen into a *pièce* of wine immediately after fermentation had ceased, with the result of imparting to it a dainty and delicate flavour.'
>
> Definitely not a recipe followed today or approved by the CIVC.

miles) to the east of Épernay and in the old province of Champagne. It is said that his father's family were vineyard owners. Not much is known about his early years, but he was educated at the Jesuit College at Châlons-sur-Marne and then, in 1657, aged seventeen, he decided that he wanted to become a priest and to join the order of the Benedictine monks at the Abbey of Saint-Vanne at Verdun. He joined the order, one of the most severe in France, where life consisted of work and prayer and one meal a day. However, Pérignon seemed to take to the monastic life and in 1668 was transferred to the Abbey of Hautvillers, just outside Épernay, where he was given the title 'Dom' and became the *cellérier*, second in command only to the abbot himself. Pérignon stayed at the Abbey for another 47 years until he died in 1715. During his time there, Pérignon was responsible for increasing the size of the Abbey's vineyards from 10 hectares (25 acres) to 24 hectares (60 acres), and for introducing several new practices into both the vineyards and cellar. According to Tom Stevenson, he instigated the construction of the first underground barrel cellar. Hewn out of solid chalk, it was able to hold five hundred barrels, and maintained an even temperature, summer and winter, something we presume Pérignon thought was important. However, inducing a secondary fermentation in the bottle to make his wines sparkle was *not* one of his new practices.

Pérignon's main claim to fame lay in recognising that the different varieties of grapes being grown at the time had different properties, and that careful blending of them was important both at the time of pressing and when blending the finished wines. At this time, the practice in most vineyards was to grow red and white varieties alongside each other, probably in the belief that, by mixing them, disease levels would be lower and thus the crop larger and healthier. It is said that Pérignon would ask for samples of grapes to be brought to him from both the Abbey's own vineyards, and from the vineyards it bought grapes from, and he would then leave them outside on the window sill for them to cool down overnight and wait until the next morning, when his stomach was empty, to taste them. In this way he got to recognise which vineyards they came from, to be able to pick at the optimum time, and to plan the pressing of them according to quality. By keeping the wines of different qualities separately, Pérignon was able to maximise the quality of the final wine after the *assemblage*. In the pressing of the grapes, of whatever colour, Pérignon realised that by gentle pressing, and by taking only the first pressing, clear white juice could be made to flow from black grapes (apparently this was the first time this had been tried) and turned into white wine. Wines at the time (other than red wines) were always described as *vin paillé* or *vin gris* or sometimes *oeil de perdrix* and pure, clear white wine was something of a rarity. Pérignon also instigated processes and techniques, many of which we today would recognise as important for wine quality: pruning for quality, not quantity; harvesting in the early morning when grapes were cool; passing through the vineyard in successive '*tries*'[3] to pick only the ripest grapes; picking into small baskets; only picking sound grapes and leaving damaged and diseased bunches; pressing near the vineyards and transporting the juice to the cellars.

Some of the more fanciful claims about Pérignon stem from a book written in 1821 by Dom Grossard, previously procurator at Hautvillers Abbey, who said that Pérignon was blind and could tell one variety of grapes from another; that he had a secret way of fining wine; and that he was the first to bottle wines using corks – all untrue. Pérignon was not blind, his winemaking methods were fully documented and corks had been used, albeit in limited amounts as bottling was not common, before Pérignon's time. The most outlandish claim for Pérignon of course was that he had invented the *méthode champenoise*. In a letter to the deputy mayor of Aÿ, Grossard wrote: 'As you know, Monsieur, it was the celebrated Dom Pérignon who found the secret of making white sparkling wine.'[4] Bonal states (quite rightly in my view) that had Pérignon worked out how to add extra sugar and yeast to dry wine at bottling time in order to make it sparkle, this surely would have been top of the list of his achievements in Grossard's book. Pérignon surely knew that still white wine often started re-fermenting in the barrel when the weather warmed up in the spring following the harvest, and that these wines had a light sparkle which gave an extra dimension to their drinking. But nothing more. Other claims such as he invented the wide and shallow 'Coquard' type grape press and that he had an involvement in making glass bottles, are also untrue. These bogus claims about Pérignon's 'invention'

---

[3] The French word 'trie' means to select (as in the English use of the word in 'triage') and in viticulture means picking the bunches of grapes in stages or 'sweeps' as they ripen. Sauternes today is typically picked in 3 to 5 'tries' as the bunches reach full ripeness.
[4] World of Fine Wine, 25 March 2014. Tom Stevenson: 'Dom Pérignon Oenothèque 1966-96'.

of sparkling wine gained a wider currency when the local *Syndicat du Commerce* decided to promote it at their stand at the 1889 *Exposition Universelle* in Paris and later (1896) published a pamphlet which 'unequivocally' stated that the 'blind monk' had 'discovered Champagne by following ancient traditions'.

Champagne after Dom Pérignon

In 1718, three years after Pérignon's death, a short treatise (only 35 pages long) called *Manière de cultiver la vigne et de faire le Vin en Champagne*.[5] was published anonymously. It covered the winemaking rules and techniques in Champagne and finished with an eight-page description, with engravings, of a *gros pressoir* (large press). It was said at the time that various people had written the book, and that perhaps Pérignon himself might even have written parts of it, but the author turned out to be a canon of Rheims Cathedral, Canon Jean Godinot (1661–1749), who had worked with Pérignon and incorporated many of his ideas into the book. Godinot, who was fifty-seven at the time of writing the book, was not revealed as the author until 1763, fourteen years after his death. He also founded Rheims's first cancer hospital, created schools and was an advocate of clean drinking water and public fountains. In Place Godinot in Rheims there is a magnificent fountain erected as a memorial to him. However, why he did not want to be known as the author of the book is intriguing.

Godinot's book lays out how to plant, establish and cultivate vineyards in Champagne and includes many of the ideas that we attribute to Pérignon: pruning not to start until 14 February to avoid frost; vineyards should separate white and black grapes and replant where necessary;[6] weeds should be kept down; and vines should be tied to oak stakes which would last twenty years. He also made recom-

Dom Perignon bottle from 1935

---

[5] A second, enlarged, edition was published in 1722 and reprinted in *La Nouvelle Maison Rustique* in 1736.
[6] This was before *Phylloxera*, and therefore before grafting. Replacement vines at this time were often established by the technique of 'layering' a shoot from an adjacent vine.

mendations about harvesting and winemaking, although here the book starts wandering. He advises picking only on days when there is dew, as this 'gives the grapes an exterior bloom, that is called *azur*, and inside a coolness, which means that they do not heat up too easily, and that the wine is not coloured'. Quite what 'azur' is, is another matter. Apart from being the French for blue (and I suppose the skin of a ripe grape covered in bloom might be described as 'blue') it appears not to have another meaning. This book is also apparently the first to mention the words *vin mousseux*[7] in relation to the wines of Champagne. However, on how the bubbles got there, Godinot is less clear. He states:

> For more than twenty years, the French taste has been determined by sparkling wine, and we loved it, so to speak, to fury and we have only started to come back a little in the last three years. Sentiments have been very divided on the principles of this type of wine; some have believed that it was the strength of the drugs that were put in it which made it foam so strongly, others attributed the foam to the greenness of the wines, because most of those which foam are extremely green; others have attributed this effect to the Moon, depending on the time that the wines are bottled.
>
> The demand for frothy wines, however, has occasioned the Merchants of wine to endeavour by Art to supply the want thereof; that is to contrive, and find out experiments, to make their wines still more frothy than they would naturally be. To which purpose, they have recourse to sundry sort of drugs, and chemical preparations to affect the same, viz by mixing Allum, Spirit of Wine, and Pidgeon's Dung therein, which 'tis certain do in some measure answer the End.

Reading the above, it is quite clear that, in 1718, someone who had been intimately connected with vineyards and winemaking in the Champagne region for several decades and who had worked closely with the most famous Champenois himself, Dom Pérignon, had no clear idea of how the sparkle came about. It mentions the 'drugs' that were put into it, without saying what the drugs were; it notes that the 'merchants of wine' also use 'sundry sort of drugs' to make the wines 'still more frothy than they would naturally be', but nowhere in the book is the word *levure* (yeast) or *sucre* (sugar) used, the two substances we know today are required for

---

[7] *Vin mousseux* today is only found on sparkling wines that are made from table wines i.e. not made from official quality wines such as champagne or the various *crémant* wines from regions such as Alsace, Burgundy, Bordeaux, Die, Limoux, Jura and Loire.

a fermentation to take place. The book discusses bottling – *mis en flacon* – (and nowhere is the word *bouteille* used), saying that if you bottle up to the month of May, some *mousse* can be expected. It adds that wine bottled between the tenth to the fourteenth of the *Lune de Mars* is sure to sparkle. Exactly what caused the sparkle though was *un mystère*. The Flemish scientist Jan van Helmont had discovered carbon dioxide in around 1630 during his experiments into what we call today photosynthesis (which he also discovered).[8] Helmont called it *gas sylvestre* as it was given off during the burning of wood. He also recognised that this was the same gas that was given off during the fermentation of wine, but it would be in another 137 years, in 1857, when Louis Pasteur, following the invention of the microscope, was able to identify the cells that turned sugar into alcohol, and, in the process, gave off carbon dioxide and heat.[9]

In 1724, nine years after Pérignon's death, yet another book was written, called *Traité de la Culture des Vignes de Champagne*. This time the author was Pérignon's 'apprentice and successor' at the Abbey, Frère Pierre (his surname appears to be unknown), who had worked alongside him and, we must assume, knew most, if not all, of his working practices. Unfortunately, this book is extremely rare and only a very few copies survive. The handwritten original was discovered, belonging to Countess Gaston Chandon de Briailles and transcribed and eventually printed in 1931 by M. le Comte Paul Chandon-Moët. I have not been able to see a copy (although Épernay library has one) but Stevenson, who I assume has seen a copy, says that the book:

> fastidiously recorded Dom Pérignon's achievements, practices, and working principles. Yet nowhere in this detailed document can we find any mention of sparkling Champagne, let alone the slightest hint that it had been invented by the modern world's most famous Benedictine monk.

---

[8] The Dutch draper Antonie Philips van Leeuwenhoek had invented a simplified microscope in the 1650s in order to study the fineness of the threads in the linens and other fabrics that he was being offered. This led to him looking at other things, including the mould on beer. This was around 1670. At the time, he didn't realise that the mould was made of living organisms – yeast – but he was sure that it was important in the beer-making process. He went on to develop many lenses and is known as the 'father of microscopy'.

[9] As all winemakers know, it was Pasteur who discovered yeast in 1857, although it is interesting to note that in 1755, a whole century earlier, Samuel Johnson, in his Dictionary of the English Language, defined yeast as 'the ferment put into drink to make it work; and into bread to lighten and swell it.' I guess Pasteur was the first to scientifically *prove* that it was yeasts that were the culprits.

CHAPTER 10

Champagne starts to sparkle

However, slowly but surely, the idea that the wines of Champagne could be sparkling took root and, slowly but surely, producers started to experiment. It was only following the death of Louis XIV in 1715 that the sparkling version of champagne became drunk at all in France, mainly because the French Regent, the Duc d'Orléans, enjoyed the sparkling version of champagne and 'featured it at his nightly *petits soupers* at the Palais-Royal'. However, sparkling champagne was not universally popular and, at the end of the 1700s, 90 per cent of champagne was a pale *oeil de perdrix* still wine. Godinot also states that:

> For more than twenty years the French taste has been for sparkling wine, a love for its fury, so as to speak, although this passion has started to regress over the last three years.

This suggest that sparkling wines were starting to appear and finding favour with the French in around the late 1600s and were obviously a matter of discussion. Bonal in his *Livre d'Or du Champagne*, states that a:

> Champagne bottle made its official appearance on 8 March 1735, accompanied by a royal decree, which stipulated that it should in future contain a pint, of Paris measure, and must not weigh less than twenty-five ounces.

In 1735, Louis XV commissioned a picture called *Le Déjeuner du huîtres* (The Oyster Lunch) from the artist Jean-François de Troy which was to hang in one of the king's 'lesser dining rooms' at Versailles. This ensured it would only be

Le Déjeuner du huîtres by Jean-François de Troy

seen by some of the king's friends and courtiers, i.e. people of influence. The picture depicts a lively lunch with around fifteen men enjoying oysters. One of the men has just opened a bottle of champagne, and the cork can be seen flying skywards, with several of the men gazing upwards at the audacity of the occasion. This royal patronage undoubtedly helped the sparkling version of champagne gain traction. It was also promoted on its lightness and frivolity, compared to the comparative heaviness of other white, red and especially fortified wines, which were widely consumed at the time. It was, if you like to compare wines which are not really comparable, the equivalent of the boom in *Prosecco* in the British market which, in ten years, went from a little-known wine to a major 'brand' and available in almost every licenced outlet in the country. *Prosecco* was certainly not promoted on how good it tastes, but on the sense of occasion that opening a bottle of (even low quality) sparkling wine brings with it.

However, both types of champagne continued to be made and sold. I have a wine list from a Bordeaux merchant dated 1760 which lists both still (*non mousseux*) and sparkling (*mousseux*) champagne (at the same price). As fashions changed and as wine bottles became more able to withstand the pressures of the secondary fermentation due to better designs and better production methods – using coal instead of timber – sparkling champagne became a much more reliable and consistent product. This enabled the larger champagne houses to market their wines overseas, as their wines did not deteriorate so quickly owing to the preservative effect of the carbon dioxide trapped in the wine and in the bottle.

In 1775, Sir Edward Barry, a Fellow of both the Royal College of Physicians and of the Royal Society, wrote: 'for some years the French and English have been particularly fond of the sparkling, frothy Champaigns [sic]. The former have almost entirely quitted that depraved taste; nor does it now much prevail here'. Barry goes on to warn against drinking wines that have: 'active gas, so powerfully injurious to the nervous system and says that those: that have indulged themselves too freely, in the use of these Wines, are particularly affected with a tremor in the nerves and spasmodic rheumatic pains'. Barry also quoted his friend Charles Hamilton, owner of the vineyard at Painshill Place, who said about his winemaking techniques: 'the only art I ever used was putting three pounds of white sugar-candy to some of the hogsheads... in order to conform to a rage that prevailed to drink none but very sweet Champaign'.

In *A History and Description of Modern Wines*, published in 1833, Cyrus Redding wrote about the problems associated with the breakage of bottles – *la casse* as it was known – in Champagne. He says that the bottles: 'are jingled together in pairs, one against the other, and those which crack, or break, are carried in account against the maker'. This however, was by no means a foolproof test of a bottle's integrity and despite inspections for air bubbles and obvious cracks, many of the bottles broke while undergoing secondary fermentation and maturation. Redding writes at some length about the precautions the workers have to go to in order to protect themselves against flying glass and states that it is normal for between 4 per cent and 10 per cent of bottles to break in store, although: 'sometimes, however, it amounts to thirty and forty per cent'. During the spring and summer (when bottles were likely to explode) visitors to champagne cellars were issued with metal masks![10]

Although Redding devotes a whole chapter to Champagne – some 24 pages going into great detail about hectarages of vines, the amount of wine made in the region, the varieties used, the exact processes involved with the initial fermentation, the way the wines were bottled, had their corks inserted and wired on, etc. – he is genuinely perplexed by the way in which the sparkle is created! He says that it is a result of the 'carbonic acid gas [produced] in the process of fermentation' and that this is due to the 'saccharine [becoming] decomposed'. This shows that he was aware of the basic principle – but the idea that this could be controlled and provoked at will by the addition of extra sugar and yeast at bottling (one of the cornerstones of the modern *méthode champenoise*), seems to have escaped him.

Was it that the French themselves did not appreciate this? I suspect so. Redding states with some authority that 'the effervescence of the Champagne wine, considered in all its bearings, is most uncertain and changeable, even in the hands of those best acquainted, through experience, with its management.' He continues, and says that there are many factors that 'all have a varied and often inexplicable influence on the phenomena of effervescence.' One wonders who it was that first fully appreciated that an addition of 24 grams of sugar per litre to a bone-dry wine, plus some active yeast, gave the required degree of carbon dioxide pressure (5–6 atm or 75–90 lbs per square inch), together with an increase of around 1.25 per cent alcohol?

---

[10] In fact, *la casse* could account for a much higher percentage of breakages and when too high an addition of sugar and sub-standard bottles met each other, losses could exceed 90 per cent of the batch.

Redding was not alone in his ignorance of the facts about the secondary fermentation. David Booth, who wrote a highly respected book called *The Art of Winemaking in all its Branches* published a year later, in 1834, was equally ignorant about why some wines sparkle and others do not. Indeed, he states quite categorically that 'the theory of fermentation, as laid down by Chaptal, is of little value to practical men'! (In 1807 Chaptal had proposed that sugar was the necessary ingredient for fermentation to take place. In fact, it wasn't until Louis Pasteur started on the problem in the late 1850s, and proposed that it was yeasts that were the real culprits, that fermentation was properly explained.)

This ignorance about the relationship between sugar, yeast and the sparkle in champagnes continued to perplex wine writers. W. H. Roberts, writing in the 5th edition of his *British Wine-Maker and Domestic Brewer* in 1849 quotes a Dr Shannan (his name was actually Shannon) who says that 'for about twenty years last past, the gust [taste] of the French has been determined for a frothy wine' and goes on to wonder about how the 'froth' gets into the wine. He writes 'some believe, that it proceeds from the force of the drugs that they [the French] put in it, which makes it froth so strongly'. Later he dismisses this possibility and puts it down entirely to inexplicable natural circumstances which are all to do with the time of bottling. He states with some certainty that 'one may always be sure to have the wine perfectly frothy, when it is bottled from the 10th to the 15th of the month of March [in the year after the harvest].' Most of this text by Roberts is, as readers will recognise, a quotation lifted (without attribution) from Godinot's 1718 book, *Manière de cultiver la vigne et de faire le Vin en Champagne*.

Around the middle of the 1800s, knowledge about the precise relationship between sugar and the sparkle in sparkling wine started to change. 'A thin brochure' that appeared in 1837, written by a Mr François, a former pharmacist in Châlons-sur-Marne, and called a *Treatise on the Preparation of Sparkling White Wines* was apparently well known to the wine merchants and dealers in Champagne. The website of the *Grandes Marques & Maisons de Champagne* states that this publication was:

> the initial and vital link in the series of trials and studies that would lead to a proper understanding of the phenomenon of how wine becomes sparkling and ultimately, as a result, the practical applications of this understanding, how to control the sparkling process and virtually eliminate the risk of breakages.

Thudichum and Dupré's *Complete manual of viticulture and Œnology* of 1872 has a long section in it about the winemaking techniques used in Champagne, describing the bottling in some detail, but covering the question of how the sparkle gets into the bottle in a few lines. They state that:

> as the Rhenish wines after their fermentation and ageing no longer contain any sugar, it is necessary to add an amount of sugar so that the whole of the sugar in the wine to be fermented is 2%. This presence of 2% of sugar, the manufacturer of Champagne mostly secures by mixing young wines only.

Even though the authors of this book go into some detail about the exact amount of sugar required to provide the required amount of *mousse* (so called, they say, because when it foams from the bottle it resembles a little patch of moss), and advise the use of a 'Schintz's Manometer' to test when fermenting bottles of champagne are nearing their breaking point due to excess pressures; it is clear that the system of champagne manufacture was, until relatively recently, very haphazard and relied upon chance to a great degree.

The system seems to have been that the young wines, i.e. those from the previous year, were allowed to ferment as slowly as possible so that some of them would stop fermenting in the cold of the winter and some of the original natural sugar, together with whatever was added at harvest time, remained in the wine. By March, when the weather might be expected to start warming up, these sweeter, partially fermented wines were blended with dry wines, whether from the previous – or even older – vintages, so that, at bottling, some residual sugar was retained in order for the secondary fermentation to take place. This ensured that the wines sparkled, although not as consistently, or probably as much, as they do today.

Glass bottle development

The change from still to sparkling happened when glass bottles which were strong and robust enough to take the pressure of a secondary fermentation became more readily available. We know that between 1630 or thereabouts and 1662, bottle-making in England developed quickly with the introduction of coal-fired furnaces and furnaces with better flues, better draught arrangements and better *lehrs* for annealing. Annealing is, after all, the part of the glassmaking process that strengthens glass of all types and shapes. We also know that bottles made of *verre Anglais* were listed in the London Port Books of 1634 as export cargoes. It is therefore probable that glassmakers working in France, whether French or Huguenots from the Low Countries, would have realised that wine could be a good market to make bottles for. In addition, wine merchants, who continued to trade in wine in barrel, would have also realised that if you could sell wine in a bottle, the number of customers would increase greatly as not everyone had the facilities (or the butlers) to take barrels into their houses and use wine out of the barrel. Geoffrey Luff, a bottle collector and enthusiast, on his website www.thevenotte.com, states that:

> On the 31st of January 1709 Gaspard Thévenot obtained a patent from the Duc d'Orleans to make bottles '*à la manière d'Angleterre*' at Folembray, today in the '*departement de l'Aisne*'. There, in the early years of the 18th century Gaspard Thévenot invented or perfected a type of bottle that became so popular that it soon carried his name. The '*Thévenotte*' bottle was born.

The *Thévenotte* bottle, similar to today's Benedictine bottles, had a squat base and lower half, said to resemble a flower-pot, with a tall tapering neck almost the same height as the lower half. Luff continues by saying that 'By 1720 France had 4 coal-fired glasshouses, by 1740 at least 14, and by 1790 about 45. These bottles were also one of the first attempts at standardization by using a simple flower-pot-shaped

Chapter 10

'dip mould' into which the body of the bottle was blown. The whole process of manufacture at the Sevres glasshouse is very well illustrated in *l'Encyclopédie de Diderot et d'Allembert* published between 1751 and 1765. The *Encyclopédie* also contains a drawing of a very fine substantial coal-fired *Verrerie Anglaise* with a tall brick-built chimney which contrasts with the much less industrial, wood-fired French glasshouses of the day (see page 60).

The transition from wood-fired to coal-fired furnaces in France changed the colour, the strength and the quality of the glass, as it had done in England over a century before. This meant that robust and equal-sized bottles, useable by merchants for both bottling and selling wine, became a reality. It is no coincidence that the eighteenth century saw the establishment of some of today's most important champagne houses: Ruinart in 1729, Moët in 1743, Veuve Clicquot, in 1772 and Roederer in 1776. Claude Moët recorded the first sale of sparkling wine in bottle in April 1744 when he sent some to Maréchal Duc de Noailles in Douai, and it would appear that the sale of sparkling champagne in bottles really started at this time. There had also been a restriction on the sale of wine in bottle in 1676 for tax reasons, but over the years this appears to have been ignored (or at least got around) and in a Decree of the King's Council of State of the 25 May 1728, 'His Majesty wishing...to encourage the Trade and Transport of the Vin Gris of Champagne' allowed wines in bottle to be transported and traded, but, strangely, 'only in baskets of fifty or one hundred bottles'.

The London auctioneers Christie's, founded in 1766, conducted its first wine sale in October 1769 which included '36 dozen of Champaigne' which averaged 30 shillings per 3 dozen bottles. Whether it was still or sparkling, or French bottled or English bottled, history does not recount, but it was definitely bottled. Another Christie's sale, this time in 1784, included *L'Oeil de Perdrix Celleroy* (Celleroy being an alternative spelling of Sillery), which is what today we might now call pink or rosé

Early Benedictine bottle

champagne, which sold for 48 shillings (£2.40) a dozen, whereas 'Ordinary Celleroy' only fetched 38 shillings (£1.90) a dozen. Whether the *Celleroy* was still or sparkling is not known but, as before, they were in bottle, although French bottled or London bottled, again, it is not known. In the same sale, a considerable amount of named First Growths were auctioned, but this time they were all in barrel and now 'in high order for Bottling'.

Charles de Saint-Denis, Seigneur de Saint-Évremond

Charles de Saint-Évremond

Saint-Évremond (1613–1703)[11,12] was a French soldier, essayist and poet, gourmet and man-about-town. Saint-Évremond was born in Normandy and after school and college in Paris, where he studied law, he joined the army and fought with distinction in the Thirty Years War (1618–1648) which was fought between France and their Swedish allies against the Spanish and Austrian Hapsburg empire. He was a founder member of the *L'Ordre de Coteaux*, which was an upmarket wining and dining club. The Bishop of Le Mans, Lavardin and himself a great gourmet, was one day at dinner with Saint-Évremond, and began to criticise him and his dining companions, the Marquis de Bois Dauphin and the Comte d'Olonne, saying:

---

[11] Claude Taittinger wrote a biography of the man, called *Saint-Evremond, ou, le bon usage des plaisirs*, which was published in 1990. Champagne Taittinger used to have a brand called Saint-Evremond, but this brand name has now been allocated to their English Sparkling Wine estate, Domaine Evremond, in Kent.
[12] His full name was Charles de Marguetel de Saint-Denis, Seigneur de St Évremond.

These gentlemen, in seeking refinement in everything, carry it to extremes. They can only eat Normandy veal; their partridges must come from Auvergne, and their rabbits from La Roche Guyon, or from Versin; they are no less particular as to fruit; and as to wine, they can only drink that of the good *coteaux* of Aÿ, Hautvillers, and Avenay.

Many of these gourmet diners were members of the Order of the Holy Spirit, one of France's oldest and noblest awards, whose cross of membership was hung on a blue riband. Because of this, the type and style of food they insisted on being served became known as *cordon bleu* cooking. Saint-Évremond was associated with Nicolas Fouquet, who was in charge of Louis XIV's finances, but who fell out of favour in 1661 and was imprisoned. In the same year, Saint-Évremond fled France, firstly to the Netherlands, but then to England, where he was received by Charles II who made him Governor of Duck Island, the small island at the Downing Street end of the lake in St James's Park, which came with a £300 a year pension.[13] He died in 1703 and is buried in Poet's Corner in Westminster Abbey, a rare honour for someone not born in Britain.

Although Saint-Évremond was a writer (mainly of poetry), he never authorised any of his work to be published during his lifetime and appears to have spent much

Bird-Keeper's Lodge in St James's Park, London

---

[13] The Royal Parks website states that the island was created in 1665 on the site of what had been a 'duck decoy', which was presumably for shooting ducks, rather than preserving them and other species of waterfowl, which is its use today.

of his time in the company of Hortense Mancini who 'set up a salon for love-making, gambling and witty conversation'. Mancini had left her marriage to one of Europe's richest men, the Duc de la Meilleraye, and fled to England in 1675, leaving behind her small children. She travelled to London under the pretext of visiting her niece, Mary of Modena, who had just married Charles II's brother, James, Duke of York, who would become James II of England (and James VII of Scotland) in 1685. Mancini allegedly travelled dressed as a man and was said to be bi-sexual. Once she arrived in London, she set out to displace Louise de Kérouaille, Duchess of Portsmouth, who was Charles II's mistress. This she achieved in 1676, gaining a pension of £4,000 a year from the King in the process. Following the death of Charles II in 1685, Mancini held court at Montagu House, the London home of the Duke of Montagu, which then occupied the site of today's British Museum (and became the home of the museum in 1753) in Bloomsbury. Here, Mancini entertained some of the best and brightest minds of the day and it was here that de Saint-Évremond appears to have been at home. It may have helped that Montagu was paying de Saint-Évremond a stipend of £100 a year which continued until his death aged ninety in 1703. Mancini lived in what was then called Paradise Row in Chelsea, but is today Royal Hospital Road (the hospital was completed in 1691) and held musical soirées which featured 'intimate musical scenes' with musicians, many of whom had also fled France. It was whilst attending Mancini's salons in Bloomsbury and Chelsea that Saint-Évremond is said to have introduced the wines of Champagne to the nobility and fashionable men and women in London and helped promote its sale.[14] It is even said that he was involved with its importation, but that maybe apocryphal and there appears to be no evidence to support that claim. Saint-Évremond, writing in 1674, said:

> The wines of the Champagne are the best. Do not keep those of Aÿ too long; do not begin those of Reims too soon. Cold weather preserves the spirit of the River wines, hot removes the *goût de terroir* from those of the Mountain.

At this time champagne was a still wine, not sparkling, and imported into Britain in barrels, as bottling wines did not start in France until later. It is also said that Saint-Évremond was horrified to find that some of the vintners and wine coopers were

---

[14] Hortense Mancini died in 1699. Before she died, she wrote her autobiography in order to get her side of the story down before her ex-husband did. The Duke of Montagu paid to have her body shipped back to France, but her ex-husband 'carted her body around with him on his travels' before she was allowed to be interred in her uncle's tomb. Today's royal family of Monaco, the Grimaldis, are descended from her.

adding molasses and sugar to still champagne and then bottling it in robust *verre Anglais* bottles in order to give it a little sparkle. However, in time he came to appreciate these sparkling champagnes and before he died described the ideal wine as: 'neither too flat nor too bubbly. It should be slightly *cremant* with a natural froth which decorates the surface with tiny bubbles.' Henry Vizetelly, in his 1882 *A History of Champagne*, describes Saint-Évremond as the *arbiter elegantiarum*[15] to the court of Charles II and he is said to have introduced the glass called the *flûte* for drinking champagne.[16]

The poet, Samuel Butler, in his poem 'Hudibras' written in 1663, calls for 'brisk Champagne' – brisk being the term used at the time for a wine that was lightly sparkling. The first actual mention of the words 'champagne' and 'sparkling' in the same breath was by the playwright, Sir George Etherege, who in 1676 wrote a comedy called 'The Man of Mode, or Sir Fopling Flutter' in which appear the following lines:

'To the mall and the park,
Where we love till 'tis dark;
Then sparkling Champaigne
Puts an end to their reign;
It quickly recovers,
Poor languishing Lovers,
Makes us frolic and gay, and drowns all our sorrow;
But, alas! we relapse again on the morrow.
Let ev'ry man stand,
With his glass in his hand;
And briskly discharge at the word of command.
Here's a health to all those,
Whom to-night we depose:
Wine and beauty by turns great souls should inspire.
Present all together; and now boys give fire …

---

[15] An authority on matters of social behaviour and taste and first used in 1728.
[16] Vizetelly also states that it was Pérignon who invented this glass shape, although in fact it had been around for several decades, if not longer. See under Sir John Scudamore and the 'Chesterfield Flute' of 1650.

Blanquette de Limoux – the world's first sparkling wine?

Claims that Blanquette de Limoux was the world's first sparkling wine appear quite often, but with little evidence to support the claim. For one thing, how could you make wine sparkling if there wasn't a bottle good enough to hold the pressure? *Petillant* maybe, but *mousseux*? *Non*. The entry for Blanquette de Limoux in the Oxford Companion to Wine, 4th Edition, states:

> For centuries it [Limoux] has been devoted to the production of white wines that would sparkle naturally after a second fermentation during the spring. They became known as Blanquette de Limoux, Blanquette meaning simply 'white' in Occitan [the language of the Pays d'Oc]. Locals claim that fermentation in bottle was developed here long before it was consciously practised in champagne, dating the production of cork-stopped sparkling wines at the Abbey of St-Hilaire from 1531 (Limoux is just north of Cataluña, a natural home of the cork oak) although Stevenson casts doubt on both the date and claims that the wine was sparkling.

The *Sieur d'Arques* co-operative winery in Limoux (which was originally called the *Société Coopérative des Producteurs de Blanquette de Limoux*) gives us some more detail about the origins of this wine: 'The Lord of the region, the Sieur d'Arques, was a great lover of this sparkling wine and liked to knock back 'flasks' of blanquette to celebrate his victories.' In 1544, the accounts of the treasurer of the city of Limoux, I. Bertrand, have an entry for: 'six justes clarets for his supper and four pinctes blanquette and two claret wines for lunch and for four flasks of claret wine'. They don't say much about the blanquette being sparkling though. The website of 'French Entrée' (an estate agent selling French property) has even more information on Blanquette de Limoux. It states that:

> the monks of the Benedictine abbey of Saint-Hilaire, near Limoux in the Aude, were producing white wine in rather an unusual fashion – instead of using oak vats the wine was fermented in a glass flask, with a cork top which gave it a natural sparkle. Apparently, Dom Pérignon passed through the valley on his way north from Spain, chanced on the abbey, stole the technique and passed it off as his own to the nobles of Champagne!

Yet another fanciful tale. No bottle, no sparkling wine. That's the problem with Limoux's claim to have made the first bottle-fermented sparkling wine.

Chapter 11

Mansell and Digby

Critical to this story is the relationship between Mansell and Digby. Did they meet in October 1623 when Digby returned from Spain after the debacle of the Infanta and the marriage proposal? Mansell was apparently consulted by the Duke of Buckingham, who was in Madrid with Charles, Prince of Wales, in December 1623. On 1 March 1624 Mansell was one of the 12 Members of Parliament appointed to investigate and draw up a list of reasons for the breaking-off of the Spanish marriage negotiations. Might Mansell have consulted Digby who, after all, was actually present in Madrid and involved with the negotiations? Mansell and Digby both served on the Navy Board around 1630 and at the same time Digby was also a Governor of Trinity House, a position of some importance, which would have meant he moved in similar spheres of influence to Mansell.

James Howell and Digby

We know that Mansell's protégé, James Howell, and Digby knew each other well and were friends. We know that they met in Madrid in 1622–1623 and that Digby healed a wound that Howell had sustained helping break up a duel, by using his 'Powder of Sympathy'. Howell had spent the previous six years either managing Mansell's Broad Street glassworks or travelling abroad all over Europe and visiting the glassmaking regions, looking at techniques of glassmaking and looking for both personnel and materials for Mansell. Surely the subject of glassmaking would have been discussed between them?

Mansell and Newnham on Severn
We believe we have evidence that Mansell had an interest in a glasshouse at Newnham on Severn either around 1613 or during the reign of Charles I, which was 1625 to 1649. The earlier date is of no help to us as Digby was born in 1603, but the later dates fit the narrative very well as they include the period after Digby's return from the Scandaroon voyage in 1629 and before the death of his beloved wife Venetia in 1633.

John Colnett and the Attorney General's investigation
We have the evidence of the four glassmakers who, in 1662, swore before the Attorney General that they had worked for Digby making bottles and that Digby had also employed John Colnett, the glassmaker who was claiming that he had perfected the making of glass bottles. We also know that the Attorney General decided, after conducting 'a long and serious consideration and examination', that Digby had invented a system of making glass bottles in around 1632. Digby was still alive (he died in 1665) but we don't know if the Attorney General consulted him about Colnett's claim.

Sir John Winter and Digby
We know that Sir John Winter, landowner, industrialist and friend (and financial backer) of Charles I, owned coal mines and glassworks in Gloucestershire and that Digby and Winter knew each other. They moved in the same circles. Winter was a major supplier of both timber and iron (for cannonballs, chains and anchors) to the Navy and it is possible that Digby and Winter had dealings about these as well. At a later date (i.e. after 1632) they were respectively Queen Henrietta Maria's Chancellor and her Secretary, and although it doesn't help with the glass bottle story, it does show that they had ample opportunity to meet and discuss mutual interests.

Digby and bottle-making
Taken together, these different connections covering the worlds of high society, the court, the Navy and last, but by no means least, glassmaking, confirm that Digby certainly had the opportunity to involve himself with glass bottles and their production. The confirmation by the Attorney General that Digby had been involved with the making of glass bottles cannot be dismissed as tittle-tattle and the date given fits in exactly with other events in Digby's life.

Appendix I: Date changes

Between 1172 and 1752, the first day of the New Year was 25 March, Lady Day. That means that any date between 1 January and 24 March belonged to what we today would call the previous year. Thus, some dates, such as 17 December 1662 occur before 18 January 1662 which, had the new year started on 1 January, would of course have been in 1663. Scotland changed the start of their year in 1600. Dates were also changed, although not by so much, in 1752, when England, Wales, Ireland and some British colonies went from the Julian calendar to the Gregorian calendar, and the day after Wednesday 2 September became Thursday 14 September, thus losing 11 days. This fiddling with the dates means that sometimes dates are either given in 'old style' or 'new style' and sometimes you have no idea if they have been adjusted or not. In this book, when dates are shown thus: January 1615/16 – it means that the date, as we would refer to it today, would be January 1616, but at the time it would be called January 1615. If you have ever wondered why April 6 is the start of a new tax year in Britain, add eleven days (plus an extra day because of a Leap Year problem) onto 25 March.

Appendix II: Glossary of glassmaking terms

Annealing
The process by which glass articles are gradually cooled until they can be handled. During the *annealing* process, which takes place in a separate furnace called a *lehr*, the glass gains much of its strength.

Bit, bit gatherer
A 'bit' of glass is a small measure of molten glass that might, for instance, be the stem or the foot of a glass, the bowl having already been made. Apprentices typically started by gathering 'bits' of glass for the master glass-blowers who actually formed the bits into the finished object.

Blowpipe, blowing iron
A *blowpipe*, or *blowing iron*, is the hollow iron tube, usually around 1.2 to 1.5 m (4 to 5 feet) long, that the glass-blower uses to collect a *gather* or *gob* of *metal* and blows down to enlarge and shape the glass.

Crucible
A *crucible* is the vessel made of *fireclay* (also called *pot-clay*) where the raw materials for the glass are placed and heated. The quality of the clay used to make *crucibles* was very important, if accidents caused by split *crucibles* were to be avoided. Also known as *pots*.

Cullet
Cullet is fragments of broken glass which can make up to 40 per cent of the materials required for making glass. The use of *cullet* makes glassmaking easier and cheaper as it helps the other raw materials melt at a lower temperature. The word probably comes from *cueillette*, French for the act of gathering, picking or collecting.

Flux
Flux is a substance that helps other substances melt at lower temperatures and, in the case of glassmaking, helps the molten glass flow. Soda and potash are often used as a *fluxes* and these can be derived from plants.

Fritting
Fritting is the pre-heating of the raw materials so that they fuse together but do not melt. The *frit* can then be ground down to a powder and stored. The use of *frit* in a batch of glass speeds up the melting process.

Gather
A *gather* is an amount of molten glass that the glassmaker collects on the end of his or her *blowpipe* to begin the making of a glass object.

Glory hole
The *glory hole* is the hole in the side of the furnace where the glassmaker collects the molten glass and where the object being made can be reheated during the glassmaking process. Some glassworks would have separate *glory holes* just for reheating the glass.

Gob
A *gob* of glass, also known as a *bit* of glass, is the piece of molten glass collected from the *crucible* by the glassmaker at the start of the glassmaking process. In a team of glassmakers, there would usually be one or more *bit* gatherers.

Knock-offs
Knock-offs, also known as *moils*, are the fragments of glass removed from the top of the object being made where the *blowpipe* was attached. This is often removed by *cracking off* the top after *annealing*.

Kick-up
Another name for the *punt* of a bottle.

Lehr
Also known as a *leer*, this is the part of the furnace where the *annealing* process takes place. It can also be a separate furnace. The word comes from the German *lehren* meaning to teach.

Marver
The *marver* is a thick piece of flat stone or metal which is used by the glassblower to roll and shape the object being made. The word comes from the French *marbre* meaning marble.

Melting
Melting is the part of the glassmaking process when the maximum heat is required and the raw materials become molten and turn into glass.

Metal
Metal is the confusing name for molten glass, as it is not metallic.

Moils
See *knock-offs*.

Parison, paraison
A *parison* (from the French *paraison*) is a partially inflated piece of blown glass held on the end of the blowpipe.

Pontil
A *pontil* is a solid rod that is attached to a piece of blown glass at the bottom, usually by gluing it on with a small dab of molten glass. This allows the glassmaker to work on the neck of the object, which in the case of bottles would be to finish the neck and apply the string-rim. When the *pontil rod* was detached if often left a sharp *pontil mark* which could be rectified by pushing the base of the bottle upwards, forming what we now call the 'punt' of the bottle. The word comes from *pont* the French word for 'bridge'.

Pots
See *crucible*.

Pot-clay
Special fireproof clay used for making *crucibles* (see above). *Fireclay* is often found in coal mines and is as valuable a product as the coal itself.

Punt
The dimple in the bottom of a bottle where the *pontil rod* was attached. Also known as a *kick-up*.

Punty
Another name for the *pontil*.

String-rim
The *string-rim* is the bead of glass applied around the top of a hand-blown bottle which protects the top of the bottle and makes an anchor for the string holding down the cork, necessary in the case of sparkling wine.

Waldglas
From the German *wald* meaning forest or wood, *waldglas* is the rustic glass produced using wood-fired kilns and is typically of a light green colour. The flux used in *waldglas* is usually of forest origin, either from the wood being burnt in the kilns, or bracken and other forest plants, collected for the purpose. It was also known as 'weald-glass'.

Appendix III: Sparkling wine methods

There are many different ways of making wine (and other beverages) sparkling. Some wines, those termed *petillant* (in French) or *spritzig* (in German) i.e. very lightly sparkling, will have a relatively low level of carbon dioxide which has been retained from the first fermentation. This is typically when the fermentation is conducted at low temperatures (around 10–12 °C) which allows some of the naturally produced gas to be absorbed into the wine. This type of wine is not truly sparkling.

For properly sparkling wine, there are in fact at least six different methods of making it, including:
- Carbonation
- Tank method (Charmat or *Cuve Close* method)
- Russian Continuous method
- *Méthode rural*, *méthode ancestrale*, or *pétillant-naturel* method
- Transfer method (*transvasage*)
- Bottle fermentation, traditional method

Carbonation
The simplest method for making a product sparkling is by carbonation, where carbon dioxide is injected into a liquid. This is common for water, fizzy drinks, beer and less expensive sparkling wines. Wine is typically not aged and is sold as soon as the process is completed.

Tank, Charmat or *Cuve Close* method
The next most cost-effective method is what is known as the *'Charmat'* method (or 'tank' or 'bulk' method). This is where the fermentation that produces the bubbles takes place in a closed pressure tank (*cuve close*) and all other downstream operations (clarifying, filtering and bottling) take place under pressure. In the case of wines such as *Moscato d'Asti* and *Asti Spumante*, the fizz is the result of a single fermentation and little or no ageing takes place so that the fruity quality of the grape juice is retained in the final wine. In the case of wines such as *Prosecco* and *Sekt*, a still 'base wine' is produced in the normal manner, but it then undergoes a secondary fermentation

with additional sugar and yeast added in a tank under pressure. Some tank-method sparkling wines can also undergo ageing in the tank and can be of high quality, although this method is mainly used for cheaper wines. *Prosecco* and *Sekt* can also be bottle-fermented and lees-aged in the traditional way, although, as this is a more expensive process, it is less commonly used. The method is named after Frenchman Eugène Charmat, who in 1907 improved and patented this process which had already been developed and patented in 1895 by Federico Martinotti (1860–1924). Sometimes also known as *metodo Martinotti,* and *metodo Italiano.*

Russian Continuous method

This is a method which makes wine sparkling via a system of tanks set up in parallel where the secondary fermentation take place. It was developed in Russia as a low-cost way of making sparkling wine or *Sovetskoye Shampanskoye* as it was called. The Portuguese wine 'Lancers' is the best-known sparkling wine made by this method. It is not widely used today.

Méthode rural, méthode ancestrale, or *pétillant-naturel* methods

After the tank methods detailed above, there are a variety of bottle-fermented techniques. Firstly, there is the *méthode rural* or *méthode ancestrale,* now sometimes called *pétillant-naturel* (or pet-nat), which is where the fermentation gets going in a tank and before it gets too advanced, i.e. before all the sugar is used up, the wine is bottled into sparkling wine bottles and the fermentation is allowed to finish in the bottle. Some producers will then riddle and disgorge the bottles (to remove the yeast lees) and at that point they can add *dosage* (sweetened wine) to the bottle before corking, muzzling, labelling, packing and then – selling. Some producers, especially those at the 'natural wine' end of the spectrum, will sell their wines exactly as they are, complete with lees and without any sweetening *dosage*. These wines really are 'naturally sparkling'. *Blanquette de Limoux* is an example of this type of wine.

Méthode champenoise, méthode traditionelle or traditional method

This method is the true *méthode champenoise, méthode traditionelle* or traditional method. This is where the secondary fermentation with additional sugar (around 24 grammes per litre (1 ounce per 1.8 pints)) and yeast takes place in a bottle, the wine remains in that bottle during the ageing process (which can be for a very long time) and after riddling, disgorging and *dosage,* the wine is sold in the same bottle.

Transfer method (*transvasage*)

There are several variations on the true 'secondary fermentation in a bottle' method. Firstly, there is the 'transfer method' where the wine undergoes the secondary fermentation with additional sugar and yeast in 'a' bottle (often a magnum size, i.e. 1.5 litres (2.6 imperial pints)) and where the wine is removed from the original bottle and transferred under pressure into a tank where it is clarified and filtered, has sweetening *dosage* added and is then put into the bottle it is to be sold in. This system used to be quite common for New World less expensive sparkling wines, but is less often found today.

A slight variation on the transfer method is common (in Champagne and worldwide) for small sizes – quarter- and half-bottles – and most bottles bigger than magnums (two normal bottles) or Jéroboams (four normal bottles) and is known as *transversage*. This is where the wine undergoes the secondary fermentation in a normal, i.e. 75 cl-sized, bottle and then, at disgorging, is decanted into the size required. This system is used because secondary fermentations in small sizes can be very variable and lead to inconsistencies in quality, and for larger sizes (which can go up to a Nebuchadnezzar holding 20 normal bottles' worth) it is for economic reasons as these bottles are sold far less often. A few champagne producers pride themselves on using half-bottles and some larger sizes for the secondary fermentation and do not use *transversage*.

Appendix IV:
Digby's 'Powder of Sympathy'

During his life, the Powder of Sympathy was one of the things that Digby was best known for. He tirelessly promoted, talked about it, and used it on others, although seemingly without gaining any material advantage from its sale and use. In 1661–1662, whilst in Florence, Digby claimed to have met a Carmelite Friar who had brought the recipe for the powder 'from the East', although this has been disputed. Some claim, with good reason, that this powder was the invention of Sir Gilbert Talbot and it is interesting to note that Digby never tried to discuss it at meetings of the Royal Society, even though he had ample opportunity to do so. Elizabeth Hedrick, writing in *The British Journal for the History of Science*[1] says:

> Sir Kenelm Digby's *A Late Discourse... Touching the Cure of Wounds by the Powder of Sympathy* (1658) is usually read in the context of seventeenth-century explanations of the weapon-salve. The salve supposedly worked by being applied to the weapon that made a wound rather than to the wound itself. But Digby's essay was as much an effort to claim priority for a powdered version of the sympathetic cure as an explanation of how the cure worked. A close examination of Digby's claims in the *Late Discourse* in the context of his own earlier work and of works by his contemporaries shows his priority claim to have been false. It was recognized as such by his most knowledgeable associates. The story of Digby's fabrications offers a case study of the generic and rhetorical terms in which seventeenth-century English thinkers made and challenged natural-philosophical claims.

The late Selim (Sam) Mellick, a vascular surgeon at the University of Queensland hospital, who wrote a mini-biography of Digby, wrote the following piece about the Powder of Sympathy.

---

[1] Vol. 41, No. 2 (Jun., 2008), pp. 161–185

Digby gained widespread recognition on the continent as well as at home by his promotion of the 'powder of sympathy' in wound management, after he presented an account of its use and a suggested mode of its action to the Academy of Sciences at the University of Montpellier in France in 1657. Digby was there for treatment of his bladder stone, and his 'discourse' on the subject was published in Paris and in London in 1658 and was reprinted more than 20 times.

His technique was to mix a salve of ferrous sulphate with discharge from a wound, and then apply the salve to the causative agent, the sword or the gun. The wound itself was simply cleaned without the usual application of caustics or cautery or other irritants, the same gentle form of wound treatment promoted by Ambrose Paré over 100 years before. Digby considered that the 'powder' worked by the transfer of 'blood atoms' from the wound coming into contact with 'curative atoms' in the salve and these were then attracted back to the wound where they effected the cure, whereas in reality, the unhindered *vis medicatrix naturae*[2] of the blandly treated wound was allowed to operate successfully.

The healing of James Howell in 1623–1624

Digby himself wrote the following account of the healing of James Howell.

> It was my chance to be lodged hard by him; and four or five days after, as I was making myself ready, he [Mr Howell] came to my House, and prayed me to view his wounds; for I understand, said he, that you have extraordinary remedies upon such occasions, and my Surgeons apprehend some fear, that it may grow to a Gangrene, and so the hand must be cut off ... I asked him then for anything that had the blood upon it, so he presently sent for his Garter, wherewith his hand was first bound: and having called for a Bason of water, as if I would wash my hands; I took an handful of Powder of Vitrol, which I had in my study, and presently dissolved it.
>
> As soon as the bloody garter was brought me, I put it within the Bason, observing in the interim what Mr Howell did, who stood talking with a Gentleman in the corner of my Chamber, not regarding at all what I was doing: but he started suddenly, as if he had found some strange alteration in

---

[2] *Vis medicatrix naturae* is an ancient medical principle and means the innate ability of the body to heal itself.

himself; I asked him what he ailed? I know not what ailes me, but I find that I feel no more pain, methinks that a pleasing kind of freshnesse, as it were a wet cold Napkin did spread over my hand, which hath taken away the inflammation that tormented me before; I replied, since that you feel already so good an effect of my medicament, I advise you to cast away all your Plaisters, only keep the wound clean, and in a moderate temper 'twixt heat and cold'.

This was presently reported to the Duke of Buckingham, and a little after to the King [James I], who were both very curious to know the issue of the businesse, which was, that after dinner I took the garter out of the water, and put it to dry before a great fire; it was scarce dry, but Mr Howell's servant came running [and told me], that his Master felt as much burning as ever he had done, if not more, for the heat was such, as if his hand were betwixt coales of fire: I answered, that although that had happened at present, yet he should find ease in a short time; for I knew the reason of this new accident, and I would provide accordingly, for his Master should be free from that inflammation, it may be, before he could possibly return unto him: but in case he found no ease, I wished him to come presently back again, if not he might forbear coming. Thereupon he went, and at the instant I did put again the garter into the water; thereupon he found his Master without any pain at all. To be brief, there was no sense of pain afterward: but within five or six dayes the wounds were cicatrized, and entirely healed.

Bibliography

Blagg, Dr Michelle, *Dr Christopher Merrett – a 17thC Man of Science*, The Alchemist, Issue 86, August 2017.

Bonal, François, *Livre d'Or du Champagne*, Éditions du Grand-Pont, Paris 1984.

Bossche, van Den, Willy, *Antique Glass Bottles, their History and its Evolution 1550–1850*, Antique Collectors' Club, Suffolk, 2001.

Chapman, Jim, *The Cider Industry and the Glass Bottle*, Gloucestershire Society for Industrial Archaeology Journal for 2012 pages 36–40. 2012

Clark, Colin Jeremy, *The glass industry in the woodland economy of the Weald*, PhD Thesis, University of Sheffield, 2006.

Cornick, Martin and Kelly, Debra (Editors), *A history of the French in London*, Institute of Historical Research, University of London, London, 2013.

Crossley, David, *The English glassmaker and his search for raw materials in the 16th and 17th centuries*, University of Sheffield, 1998.

Crowden, James, *Cider – The Forgotten Miracle*, Cider Press 2, Somerset, 1999.

Diderot, D'Alembert, *L'Encyclopédie Diderot et D'Alembert – L'art du Verre*. Paris, 1750–72

Digby, Sir Kenelm, *The Closet of Sir Kenelm Digby Knight Opened*, Digby, John 1669. Edited by Jane Stevenson and Peter Davidson, Prospect Books, Totnes, Devon, 2010.

Digby, Sir Kenelm, *The private memoirs of Sir Kenelm Digby*, Saunders and Otley, London, 1827.

Dumbrell, Roger, *Understanding Antique Wine Bottles*, Antique Collectors Club, 1983.

Ellis, Jason, *Glassmakers of Stourbridge and Dudley 1612–2002*, Jason Ellis, Harrogate, 2002.

Eyre, Hermione, *Viper Wine*, Jonathan Cape, London, 2014.

Faith, Nicholas, *The story of Champagne*, Hamish Hamilton, London 1988, and Infinite Ideas, Oxford 2016.

Fielden, Christopher, *Is this the wine your ordered Sir?* Christopher Helm Ltd, Bromley, 1989.

Forbes, Patrick, *Champagne, the wine, the land and the people*, Victor Gollancz, London, 1989.

BIBLIOGRAPHY

French, Richard K., *The history and virtues of cyder*, Robert Hale Ltd, London, 1982.

French, Richard. V., *Nineteen Centuries of Drink in England: A History*, Second Edition – Enlarged and Revised, London, National Temperance Publication Depot, 2014.

Gladstone, Sir Hugh, *Dr Christopher Merret and his Pinax Rerum Naturalium Britannicarum*, British Birds Rarities Committee website. www.bbrc.org.uk

Godfrey, Eleonor S., *The Development of English Glassmaking 1560–1640*, Oxford University Press, Oxford, 1975.

Godinot, Jean, *Manière de cultiver la vigne et de faire le vin en Champagne*, Savoirs Et Traditions, Paris, 1718.

Guy, Kolleen M., *When Champagne became French*, John Hopkins University Press, Baltimore and London, 2003.

Henderson, Alexander, *The History of Ancient and Modern Wines*, Baldwin, Cradock and Joy, London, 1824

Hart, Cyril, *The Industrial History of the Dean*, David & Charles, Exeter, 1971.

Hartshorne, Albert, *Old English Glasses*, Edward Arnold, 1987.

Hill, Christopher, *God's Englishman, Oliver Cromwell and the English Revolution*, Weidenfeld and Nicholson, London, 1970.

Jeffreys, Henry, *The Empire of Booze*, Unbound, London 2017.

Kenyon, G. H., *The glass industry of the Weald*, Leicester University Press, 1967.

Longueville, Thomas, *The Life of Sir Kenelm Digby*, Longmans, Green, and Co, London 1896.

Louis-Perrier, M., *Mémoire sur le vin de Champagne*, Bonnedame Fils, Épernay, 1886

Ludington, Charles C., *Inventing Grand Cru Claret: Irish Wine Merchants in Eighteenth-Century Bordeaux.* Global Food History, 5:1–2 2019

Macdonell, Anne, *The Closet of Sir Kenelm Digby Knight Opened*, Philp Lee Warner, London, 1910.

Markham, Gervase, *The English Housewife*, London, 1615.

Mauck, Aaron, *By Merit Raised to That Bad Eminence: Christopher Merrett, Artisanal Knowledge, and Professional Reform in Restoration London*, Medical History 56: 27–47, 2012

Mellick, Sam A., *Sir Kenelm Digby*, ANZ Journal of Surgery, Royal Australasian College of Surgeons, 2011.

Meredith, John, *The Iron Industry of the Forest of Dean*, The History Press Ltd, Cheltenham, 2006.

Merret, Christopher, Dr., *L'Arte Vetraria* (The Art of Glass), translation with additional notes published in 1662. Original written in 1611.

Moshenska, Joe, *A Stain in the Blood: The Remarkable Voyage of Sir Kenelm Digby*, Heinemann, 2017.

Morgan, Kathleen and Smith, Brian S, *Newnham: Economic history* in *A History of the County of Gloucester: Volume 10*, Victoria County History, London, 1972.

Munk, William, *The roll of the Royal College of Physicians 1518 to 1700s*, Royal College of Physicians, London, 1878.

Munk, William (Editor, Fine, Leon G), *Harvey's Keepers, Harveian Librarians Through the Ages*, Royal College of Physicians, 2007.

Parker, Thomas, *Tasting French Terroir*, University of California Press, Oakland, 2015

Penn, Herbert, *Glassmaking in Gloucestershire*, Gloucestershire Society of Industrial Archaeology Journal, 1983, p 3–16.

Paynter, Sarah, *The importance of pots: The role of refractories in the development of the English glass industry during the 16th and 17th centuries*. Annals of the 18th Congress of the International Association for the History of Glass, Corning Museum of Glass, Corning, New York, 2009.

Phelps, Matt, *Shinrone Glasshouse, Co Offaly, Ireland. Analysis of the 17th Century Glass Vessel Fragments*, English Heritage, Research Department Report No. 97, 2010.

Pierre, Frère, *Traité de la Culture des Vignes de Champagne*, M. le Comte Paul Chandon-Moët, 1931.

Powell, Harry J., *Glass-Making in England*, Cambridge University Press, 1923.

Robinson, Andrew, *Vintage Years*, Phillimore and Co Ltd, Chichester, 2008.

Simon, André, L, *The History of Champagne*, Ebury Press, London, 1962.

Simon, André, L, *The History of the Wine Trade in England*, Wyman and Sons, 1906–7.

Stevenson, Tom, *Christie's World Encyclopaedia of Champagne & Sparkling Wine*, Absolute Press, London, 1998.

Thrush, Andrew Derek, *The Navy under Charles 1, 1625–40*, University College London, PhD Dissertation 1990.

Tyler, Kieron and Willmott, Hugh, *John Baker's late 17th-century glasshouse at Vauxhall*, Museum of London Archaeology Service, Monograph 28, 2005.

Unwin, Dr Tim, *Wine and the Vine – An Historical Geography of Viticulture and the Wine Trade*, London, Routledge, 1991.

Vizetelly, Henry, *A History of Champagne*, Vizetelly and Co, London, 1882.

Walters, Robert, *Bursting Bubbles, a secret history of Champagne and the rise of the great growers*. Quiller Publishing, London, 2018.

Wheatley, Henry B, *London Past and Present: Its History, Associations, and Traditions Volume III*, John Murray, London, 1891

Woods, Mabel K., *Newnham-on-Severn*, Longmans, Green and Company, London, 1912.

Worlidge, John, *Vinetum Brittanicum, or a Treatise on Cider*, London, 1676.

Image acknowledgements

Page	Description	Copyright owner
Front cover	Sir Kenelm Digby	National Portrait Gallery, London
10	Sir Kenelm Digby	National Maritime Museum, Greenwich, London.
11	Charles I	Stephen Walli
12	Chesterfield Flute	Museum of London
19	Bottle with seal	Rijksmuseum, Amsterdam
23	Chapel Down	English Wines PLC
27	Old wine press	Chris Lake, Southern Oregon Wine Institute
30	Cellars	Harper -1040
33	Amphorae	Ed Meskens
34	Cinque ports boat	Dover Library
43	Bottling	The Wine Society
46	Isaac Judeus	Rijksmuseum, Amsterdam
46	Shabti	Metropolitan Museum of Art, New York
47	Roman bottle	Metropolitan Museum of Art, New York
49	Marver	Voltage TV - Inside the Factory
50	Bottle	Rijksmuseum, Amsterdam
52	Shinrone	Paul Francis
55	Coppicing	Westernbirt Arboretum, Rypelwood Workers Co-Op
58	Prince Royal	Rijksmuseum, Amsterdam
68	Canal	Graham Fisher MBE, Trustee British Glass Foundation
68	Red House Glass	www.encrovision.uk
72	Bottle	Voltage TV - Inside the Factory
79	Bottle seal	Ashmolean Museum
80	Bottle	Corning Museum of Glass, Corning, NY
80	Seal	Corning Museum of Glass, Corning, NY
81	Sir Kenelm Digby	National Portrait Gallery, London
94	Venetia Digby	Dulwich Picture Gallery, London
96	Gresham Place	Gresham College, London
99	Oliver Cromwell	Elliott Brown
118	Merret's paper	The Royal Society, London
125	Dom Pérignon	Michael Osmenda
131	Le Déjeuner du huîtres	Musée Condé, Collections des musées de France
135	Remuage	Gérald Garitan
160	Author	Wendy Wilmot

Notes

P.94 Anthony van Dyck, *Venetia, Lady Digby, on her Deathbed*, 1633, oil on canvas, 74.3 x 81.8 cm, DPG194. Dulwich Picture Gallery, London

P.80 *Wine Bottle* (about 1678). CMOG 54.2.14. Image licensed by The Corning Museum of Glass, Corning, NY (www.cmog.org) under CC BY-NC-SA 4.0.

Biography - Stephen Skelton MW

Stephen Skelton has been involved with growing vines and making wine since 1975. He spent two years in Germany, working at Schloss Schönborn in the Rheingau and studying at Geisenheim, the world-renowned college of winegrowing and winemaking, under the late Professor Helmut Becker. In 1977 he returned to Britain to establish the vineyards at Tenterden in Kent (now home to Britain's largest wine producer, Chapel Down Wines), and made wine there for 22 consecutive vintages. From 1988 to 1991 he was also winemaker and general manager at Lamberhurst Vineyards, at that time the largest winery in the country. During his time as a winemaker, he won the Gore-Browne Trophy, awarded for the 'English Wine of the Year', in 1981, 1990 and 1991. Stephen now works as a consultant to vineyards and wineries in the UK, setting up new vineyards for the production of both still and sparkling wine and helping existing growers expand. He was instrumental in finding the site for and setting up Domaine Evremond, the Champagne Taittinger vineyards and winery in Kent and has planted all their vineyards to date.

The author and statue of Dom Pérignon at Moët et Chandon, Épernay

In 1986 Stephen started writing about wine and lecturing for the Wine and Spirit Education Trust (WSET) and has contributed articles to many different publications. He has written several books on the wines of Great Britain: *The Vineyards of England* in 1989, *The Wines of Britain and Ireland* for Faber and Faber in 2001 (which won the André Simon Award for Drinks Book of the Year), the *UK Vineyards Guide* in 2008, 2010 and

2016, and *The Wines of Great Britain*, part of the Classic Wine Library, in 2019. In 2014 he wrote and published *Wine Growing in Great Britain* and published the 2nd Edition of this book in 2020. His best-selling book is *Viticulture – An introduction to commercial grape growing for wine production* which was first published in 2007, with an updated 2nd edition published in 2020, and has sold well over 10,000 copies. This book is aimed at WSET Diploma and Master of Wine candidates and has been translated into Japanese and Chinese. He was for many years the English and Welsh vineyards contributor to the annual wine guides written by Hugh Johnson and Oz Clarke, and currently writes the section on English and Welsh wine in both Jancis Robinson's *Oxford Companion to Wine* and Hugh Johnson and Jancis Robinson's *World Atlas of Wine*.

Stephen was a director of the English Vineyards Association (EVA) 1982–1995 and of its successor organisation, the United Kingdom Vineyards Association (UKVA) 1995–2003. He was Chairman of the UKVA 1999–2003. He was also at various times between 1982 and 1986 Treasurer, Secretary and Chairman of the South East Vineyards Association, Secretary of the Circle of Wine Writers between 1990 and 1997 and has served on various EU committees in Brussels representing UK winegrowers. Since 2018 he has chaired WineGB's Viticultural Working Group.

In 2000 he completed a BSc in Multimedia Technology and Design at Brunel University. While at Brunel, Stephen was awarded the Ede and Ravenscroft Prize for his final year project, a touch-screen 'retail wine selector'. In 2011 he was awarded an Honorary Doctor of Business Administration from Anglia Ruskin University and in 2012 he completed an MA in Creative Writing at the University of East Anglia.

In 2003 Stephen became a Master of Wine (MW), winning the prestigious Robert Mondavi Trophy for gaining the highest marks in the theory section of the examination and in 2005 won the AXA Millésimes Communicator of the Year Award for services to the MW education programme. Between 2003 and 2009 he served on the MW Education Committee and was Course Wine Coordinator, served on the Council of the Institute of Masters of Wine between 2009 and 2015 and was Chair of the Research Paper Examination panel from 2013 to 2021. He became a Freeman of the Vintner's Company in 2020.

Stephen has been the panel Chairman for English and Welsh wines for the Decanter World Wine Awards since 2008 and has judged in the past for the International Wine Challenge, the International Wine and Spirit Competition, the Japan Wine Challenge and the Veritas Wine Awards.

In 2020, Stephen was awarded the Wines of Great Britain Lifetime Achievement Award even though, at the age of seventy-two, he believes he still has a few years remaining.

Index

Indexer: Dr Laurence Errington

abbeys *see* monks and monasteries
Académie Parisienne, 99
Accolade Park, 43
Addison, Joseph, 40
adulteration, 38, 40
ale and beer, 8, 69, 76–78
Algiers, Digby in, 88
Amen Corner and Merret, 113, 114, 122
amphora, 33
annealing, 48–49, 136, 146
Antique Glass Bottles, their History and its Evolution, 72
apothecaries and medical/chemical practitioners, 52, 73
    Merret's battle against, 115–116
Art of Glass, Neri's, 10, 51, 55, 61, 74, 93
    Merret's translation, 10, 51, 55, 61, 116
Art and Mystery of Vintners and Wine Coopers, 119fn
Art of Winemaking in all its Branches, 134
Attorney General, 13, 82, 90, 91, 92, 144
Aubrey, John (diarist), 87
Aÿ, 124–125, 127, 139, 140
azur, 129

barrels, wine transport in, 37–38
Barry, Sir Edward, 132
basket wine press, 27
Beale, Rev. John, 12, 77, 78, 121
beer and ale, 8, 69, 76–78
Benedictine monks and monasteries, 126, 130, 136, 142
Bigo, Abraham, 66
bit (gob/gather), 15, 49, 53, 146, 147
bladder stones, 100, 154
Blanquette de Limoux, 142
blowpipe (blowing iron), 40, 49, 50, 146, 147
Bodicott, Humphrey (and daughter Judith), 78–79
Bodleian library, Digby's donation to, 103

Bonal's *Livre d'Or du Champagne*, 117, 125, 127, 131
Booth, David, 134
Bordeaux, 35, 37
    châteaux wines, 28, 29, 40–41, 41
    négociant system, 39–41
bottles (glass), 8, 71–78, 136–138
    beer and ale, 8, 76–78
    champagne, 131, 132, 133, 134, 135, 137
    Digby and bottle-making, 90–93, 144
    fermentation in, 7, 23–25, 126, 132, 133, 136, 142, 150–152
    pressure of, 8, 24–25, 29, 132, 133, 136
Bridgeman, Henry Bridgeman, 109
Brief Lives, 82fn, 87fn
Bristol, 43, 80, 109
Britain
    blockage of French ports (1803-1815), 29
    glassmaking, 51–58
    Romans in, 29, 33–34, 51, 53
    wine trade *see* wine trade
    see also England; Scotland
British Wine-Maker and Domestic Brewer, 134
Broad Street, 11, 66, 67, 85, 93, 103, 143
Butler, Samuel (poet), 141

carbon dioxide, 23, 24, 132, 133, 150
    discovery, 130
carbonation, 150
Carré, Jean, 56
Catholicism, Digby's conversion to, 83
    for second time, 96
Champagne (region), 8, 123, 124–125, 129
champagne (sparkling wine), 123–142
    sparkling, 131–137
        bottles, 131, 132, 133, 134, 135, 137
    still, 125, 131, 132, 140

INDEX

Champagne: The Wine, the Land, and the People, 117
Chandon-Moët, M. le Comte Paul, 130
Chapel Down, 7, 23, 160
Charles I of England, 11, 68–69, 98, 110
    as Prince of Wales (before ascending to throne), 84–85, 86
        Infanta Maria Anna and, 84–85, 104, 143
    wife (Queen Henrietta Maria), 11, 14, 85, 87, 98, 102, 110, 111, 144
    Winter (Sir John) and, 110, 144
Charles II of England, 11, 12, 14, 38, 78, 101, 105, 139, 140
Charleton, Dr Walter, 117–119
Charmat (cuve close; tank) method, 24, 150–151
châteaux wines, 28, 29, 40–41, 41
chemical practitioners *see* apothecaries and medical/chemical practitioners
Chesterfield Flute, 12, 141fn
Christie's (London auctioneers), 137–138
cider industry, bottles, 76–78
Cinque Ports, 35–36
Civil War, 14, 69, 92, 103, 105, 109, 110
    see also Cromwell
claret, 28, 40, 142
Clavell, Sir William, 66, 67
clay for crucibles (pot-clay/fireclay), 55, 59, 67, 146, 148
cleanliness and sterility, 29
Closet of the Eminently Learned Sir Kenelme Digbie Kt. Opened, 96, 102–103
Co-operative Wholesale Society (CSS), 43
coal (for furnaces), 21–22, 50–51, 57, 75, 107, 108, 136, 137
    change from timber to, 59–61
    Forest of Dean, 58, 59, 109
Coke, Sir Edward, 82
Colnett, Jean (Jean Colinet), 13, 90, 91, 144
Comley, Peter, 63
Complete manual of viticulture and Œnology, 134–135
Concerning the Different Wits of Men, 118
Constellation Park (former name of Accolade Park), 43
coppice woodlands, 55
cork (stopper), 74, 79
    sparkling wine incl. champagne, 127, 133, 142
courtiers, 40–41, 62fn, 83
Cromwell, Oliver, 11, 75, 99, 101
    see also Civil War
Crown Tavern (Oxford), 79, 80

crucibles (pots), 48, 49, 55, 59, 61, 146, 147
    clay for (pot-clay/fireclay), 55, 59, 67, 146, 148
Crutched Friars, 56–57, 75, 109
cullet, 50, 73, 75, 146
cute (Merret's use of term), 121
cuve close (Charmat; tank) method, 24, 150–151

Darby, Abraham, 111
Dark Ages *see* mediaeval times
dates and the Julian/Gregorian calendars, 145
de' Medici, Marie de', 84, 97
de Rothschild, Baron James, 41
    Baron Philippe, 41
    Nathaniel, 41fn
de Troy's painting *Le Déjeuner du huîtres*, 131–132
Déjeuner du huîtres, 131–132
Delos, Digby in, 89
de-stemming, 29
Development of English Glassmaking, 92
Digby, Sir Kenelm, 10, 75, 81–105, 143, 144, 153–155
    bottle-making, 90–93, 144
    claim as inventor of, 90–93
    Closet of the Eminently Learned Sir Kenelme Digbie Kt. Opened, 96, 102–103
    cousin George (second Earl of Bristol), 102
    cousin John (first Earl of Bristol), 83, 84, 102, 104
    death, 101
    donation to Bodleian and Harvard Libraries, 103
    exile/banishment, 97–99
    father (Sir Everard), 81–82
    Howell and, 11, 85, 93, 104, 143, 154–155
    imprisonment, 57, 97
    Mansell and, 10, 68, 75, 143
    Merret and, 10, 93, 116, 121
    Newnham on Severn and, 10, 75, 93, 107, 108, 109, 110, 144
    Powder of Sympathy *see* Powder of Sympathy
    return from exile, 99–101
    son John, 87, 96, 102
    son Kenelm, 87
    voyage to and return from Sanderoon, 88–89
    wife (Venetia Nastasia), 86–87
        death, 94–97
    Winter and, 93, 111, 144
Discourse Concerning the Vegetation of Plants (paper), 101
Discourse on Cider (paper), 13, 77
Discourse on the Powder of Sympathy (paper), 100

Dom Pérignon, 125–128
    see also Pérignon, Dom
Doomsday Books, 34
Dudley, Lord, and illegitimate son Dud, 59–60
Dumbrell, Roger, 73, 75
Dutch 'shaft and globe' bottles, 19, 50, 72, 75

Edward III, 35, 37, 109
Edward VI, 37, 56
Egypt, ancient, 46
encépagement, 40
Encyclopédie de Diderot et d'Allembert, 60, 137
England
    French areas owned by, 124
    Navy see Navy
English Housewife, 74
Epernay, 124, 126, 130
    Moët et Chandon, author at, 160
European mainland, mediaeval times, 35, 47–48, 123–124
Evelyn, John, 78, 99
Eyre, Forest (of Dean), 107, 110

Farrar's epitaph on Digby, 101
Fawkes (Guido/Guy), 81, 82
fermentation, 23–25
    primary, 24
    secondary (refermentation)
        bottles see bottles
        Merret's paper, 119–120
        methods apart from bottles, 24, 150–152
        yeast and sugar for, 7, 23, 24, 120, 127, 129–130, 133, 134, 151, 152
Fielden's *Is this the wine your ordered Sir?*, 117
fireclay (pot-clay), 55, 59, 67, 146, 148
'flat' wines (Merret's use of term), 120
Florence, Digby in, 84, 93, 153
flûte, 141
    see also Chesterfield Flute
flux, 45, 47, 48, 147, 149
Foires de Champagne, 123
Forbes' *Champagne: The Wine, the Land, and the People*, 117
Forest of Dean, 14, 58, 59, 107, 109, 110, 111
fortified wine, 38, 120, 132
Fouquet, Nicolas, 139
France, 8, 35
    British blockage of ports (1803-1815), 29

Digby, (incl. Paris), 96–97, 97–99, 99–100, 101, 154
England owning part of, 124
    see also Bordeaux; champagne; Rheims
François, Mr, 134
fritting, 50, 60, 147
fuel (for glassmaking), 21–22, 46, 48, 50–51, 54–55
    see also coal; wood
furnaces, 9, 48–49, 56–57, 61
    fuel for see coal; fuel; wood
    glory hole, 147

gather (bit/gob), 15, 49, 53, 146, 147
George Etherege, Sir George, 141
German wines, 35
glass bottles see bottles
Glass Industry of the Weald, 108
glassblowing, 46, 49, 73–74
glassmaking (in general), 45–69
    British Isles, 51–58
    developments, 54–58
    fuel changes (timber to coal), 59–61
    Mansell and, 66–69
    Merret and, 116
    see also bottles
Glassmaking in Gloucestershire, 107
glory hole, 147
Gloucester and Gloucestershire, 93, 107–110, 113
gob (bit/gather), 15, 49, 53, 146, 147
Godinot, Canon Jean, 126, 128–129, 131, 134
Grand Tour, Digby's, 84, 86
grapes
    mediaeval winemaking, 27–29
    Pérignon and varieties of, 126
    presses (wine presses), 20, 27, 28, 29, 127
    see also vineyards
Great Fire of London, 38fn, 114, 122
Great Plague, 38fn, 114
Gregorian and Julian calendars, 145
Gresham College, 77, 95, 96, 100, 101, 116, 118
Grossard, Dom, 127
Gunpowder plot, 81, 82

Hamey, Dr Baldwin, 122
Hamilton, Charles, 132–133
Hartman, George (Digby's steward), 96
Harvard library, Digby's donation to, 103
Harveian Librarian, Merret as, 113
Harvey, Dr William, 113, 114, 115

INDEX

Haut-Brion, 38, 39fn
Hautvillers Abbey, 124, 126, 127, 139
Henderson's *The History of Ancient and Modern Wines*, 117
Henrietta Maria (Queen), 11, 14, 85, 87, 98, 102, 110, 111, 144
Henry, Prince of Wales, 35, 65
Henry IV of France, 14, 85, 124
    his daughter Queen Henrietta Maria, 11, 14, 85, 87, 98, 102, 110, 111, 144
Henry IV Parts I and II (Shakespeare), 37, 38
Henry V of England, 124
Henry VIII of England, 9, 21, 83, 96
hermitaging of wines, 40
Hills Place cellars (Wine Society), 42, 43
historical timelines, 15
History and Description of Modern Wines, 133–134
History of Ancient and Modern Wines, 117
History of Champagne, 117
Holden, Henry, 90fn
Houghton, John, 108
Howard, Lord Charles, 64
Howard, Sir Thomas, 62
Howell, James, 11, 103–105, 143
    Digby and, 11, 85, 93, 104, 143, 154–155
    wound healing, 154–155
Hudibras (Bulter's poem), 141
Huguenots, 9, 48, 60, 109, 136
Hundred Years War, 35, 124
Huniades, Johannes Banfi, 96

imports of wine, 34–39
Industrial Revolution, 22
Ireland
    17th C glasshouse at Shinrone, 52, 66–67fn
    wine trade, 39
iron smelting, 21, 46, 51, 109
Is this the wine you ordered Sir?, 117
Iskenderun (Sanderoon), Digby's voyage to and return from, 88–89

James I of England (James VI of Scotland), 9, 83, 85–86
    Digby and, 10, 85, 85–86
    wood use banned by, 21, 51, 62, 63, 109
James II of England, 14, 38, 85fn, 140
Jefferson, Thomas (US President), 39
Joan of Arc, 124
John, King, 37
Julian and Gregorian calendars, 145

kali (potash), 47, 75, 147
Kenyon's *Glass Industry of the Weald*, 108
kick-up (punt), 50, 73, 147, 148
Kimmeridge, 66, 67
King's Butler, 37, 72
knock-offs (moils), 50, 147

Lafite (Chateaux), 39, 40, 41
Lambeth, 66, 67
Late Discourse, 153
Le Febure (le Fèvre), Nicholas, 98
lead crystal glass, 55–56
leer (lehr), 48–49, 136, 147
Leeuwenhoek, Antonie van, 130fn
lehr (leer), 48–49, 136, 147
Leith, 34–35, 40, 41
letters patent, 13, 21fn, 90, 91
Libourne, 35
lignite, 50
lime (calcium hydroxide), 47, 48, 75, 93
Limoux and Blanquette de Limoux, 142
Liscourt (Abraham) and family, 109
Livre d'Or du Champagne (Bonal's), 117, 125, 127, 131
London
    Great Fire of, 38fn, 114, 122
    Mansell's glassworks in, 66, 67–68
    Merret in, 113–114
    wine trade, 39–40
Louis XIII, 84, 97
Louis XIV, 84, 100, 102, 131, 139
Louis XV, 131
Luff, Geoffrey, 136
Luther, Martin, 9, 48

Madrid
    Digby in, 11, 83, 84–85, 93, 104
    Howell in, 93, 104
Mancini, Hortense, 140
Manière de cultiver la vigne et de faire le vin en Champagne, 128, 134
Mansell, Frances (Dr), 104
Mansell, Robert (Admiral Sir), 9–10, 21, 62, 63, 64–69, 143
    Digby and, 10, 68, 75, 143
    glassmaking and, 66–69
    Howell and, 104
    Newnham on Severn and, 68, 107, 108, 109, 110, 144
    patents, 62–63, 67, 69

Maria Anna, Infanta, 84–85, 104, 143
Marshalsea prison, 65, 67
Martinotti, Federico, 151
marver, 49, 148
mediaeval times (Dark Ages; Middle Ages), 34
    glassmaking, 47–48
    wine trade, 30–31, 34–37
    winemaking, 27–31
medical practitioners *see* apothecaries and medical/chemical practitioners
Mellick, Selim (Sam), 153–154
melting, 49, 148
Merret, Dr Christopher, 19, 51, 55, 77fn, 113–122
    Digby and, 10, 93
    glass and, 116
    Neri's *Art of Glass* translated to English, 10, 51, 55, 61, 116
    papers on wine, 117–121
Mersenne, Père Marin, 97
'metal' (molten glass), 49, 73, 93, 148
méthode ancestrale (rural), 150
méthode champenoise (tradionelle), 127, 133, 151
méthode rural (ancestrale), 150
méthode traditionelle (champenoise), 127, 133, 151
metodo Martinotti (*Charmat*; *cuve close*; tank) method, 24, 150–151
Middle Ages *see* mediaeval times
Milford Haven, 66, 68
Miller, Hugh, 74–75
Milos, Digby in, 89
Moët, establishment, 137
Moët et Chandon, author at, 160
moils (knock-offs), 50, 147
molten glass ('metal'), 49, 73, 93, 148
monks and monasteries and abbeys, 51, 123
    Benedictine, 126, 130, 136, 142
Mont le Ros (killed by Digby in duel), 97
Montgomery, Earl of, 62
Morgan and Smith's *Newnham: Economic History*, 108
Murano, 47, 55, 56, 103
mute/muté/mutage, 120
Mysterie of the Vintners, 117, 118

Nailsea, 80
Napoleonic Wars, 29
Navy (English)
    Digby and, 90
    Mansell and, 64–65

négociant system, 39–41
Neile (Neil), Sir Paul, 13, 77, 78
Newcastle-upon-Tyne and Mansell, 66, 67, 68, 69, 109
Newent, 109
Newnham on Severn, 80, 93, 107–110, 110, 111, 144
    Digby and, 10, 75, 93, 107, 108, 109, 110, 144
    Mansell and, 68, 107, 108, 109, 110, 144
Normans, 34, 72

Oeil de Perdrix Celleroy, 137–138
Oldenburg, Henry Oldenburg, 12, 77
Oxford
    bottles, 78–79
    Digby's donation to Bodleian library, 103
    Howell at, 103
Oyster Lunch (*Le Déjeuner du huîtres*), 131–132

Painshill Place, 132
Palmer, Sir Geoffrey, 91, 92
Paris, Digby in, 96–97, 97–99, 101, 154
parison (paraison), 49, 148
Pasteur, Louis, 130, 134
patents, 13, 21fn, 56, 57, 61–63, 109, 136
    Digby's, 90–91
    letters, 13, 21fn, 90, 91
    Mansell's, 62–63, 67, 69
    Slingsby's, 61–62, 109
Penn, Herbert, 107
Pepys, Samuel, 8, 38–39, 76, 78, 111
Percival, Edward, 13, 62, 91
Pérignon, Dom (Pierre), 6, 125–128, 129, 130, 142
    author and statue of, at Moët et Chandon, Épernay, 160
pétillant-naturel, 151
Phoenicians, 33
Pinax Rerum Naturalium Britannicarum, 121–122
plague (Great Plague), 38fn, 114
pontil (punty), 40, 50, 53, 73, 148
pot *see* crucible
pot-clay (fireclay), 55, 59, 67, 146, 148
potash, 47, 75, 147
Powder of Sympathy, 104, 143, 153–154
    Discourse on the Powder of Sympathy (paper), 100
prehistory, 23–24
presses (grape/wine), 20, 27, 28, 29, 127
pressure in bottle, 6, 24–25, 29, 132, 133, 136
private memoirs of Sir Kenelm Digby, 86, 87, 89

Index

privateer, Digby as, 88–90
'Proclamation Touching Glasses' (1615), 9, 21, 56, 63
Prosecco, 132, 150, 151
Protestantism, 9, 48
    Digby's conversion to, 83, 86
punt (kick-up), 50, 73, 147, 148
punty (pontil), 40, 50, 53, 73, 148

Red House Glass Cone, 68
red wine, 40, 42, 125
    mediaeval times, 28
Redding's *A History and Description of Modern Wines*, 133–134
refermentation *see* fermentation, secondary
Restoration of the Monarchy, 8, 11, 14, 92, 99, 111
Rheims, 123–124, 128
Roberts, W.H., 134
Romans, 29, 33–34, 47, 51, 123
    windows, 53
Rothschild *see* de Rothschild
Royal College of Physicians (RCP)
    Barry (Sir Edward) and, 132
    Merret and, 113, 114–115, 122
Royal Society, 10, 11, 12, 13, 77–78, 91, 100, 115, 116, 132, 152
    founding, 100
    Merret and papers presented to, 117–121, 122
Russian continuous method, 151

Sackville, Richard (Earl of Dorset), 87
Sadler, William, 13, 91
Saint-Évremond, Seigneur de (Charles de Saint-Denis), 138–140
Saint-Hilaire Abbey, 142
Salutation (inn), 79
sand, 45, 46, 48, 55, 75
Sanderoon, Digby's voyage to and return from, 88–89
Schintz's Manometer, 135
Scotland, wine trade, 34–35
    Leith, 34–35, 40, 41
Scott, Acton, 74–75
Scudamore, Sir John, 12
sea-coal, 51fn, 62, 63
seals on bottles, 79
seaweed, 45, 47
Sekt, 150, 151
Severn (River), 59, 108
    see also Newnham on Severn

Sevres glasshouse, 137
'shaft and globe' bottles, 19, 50, 72, 75
Shakespeare, 8, 21, 73–74
    Henry IV (Parts I and II), 37, 38
Shannan (Shannon), Dr, 134
Shinrone 17th C glasshouse, 52, 66–67fn
Sienna, Digby in, 84
Simon's *The History of Champagne*, 117
Skelton, Stephen, biography, 160–161
Slingsby, Sir William, 61–62, 109
smelting iron, 21, 46, 51, 109
Society, The (forerunner of Royal Society), 100
soda, 47, 147
Some Observations Concerning the Ordering of Wines, 10, 117, 118, 119
Southwark, 57, 62, 66, 67, 97
sparkling wines, 5–6, 23–25, 131–137
    champagne *see* champagne
    difference from still wines, 24
    methods other than bottle-fermentation, 5, 24, 150–151
    origins, 8
    world's first, 142
Stanley, Venetia Nastasia (wife of Digby) *see* Digby
steel marver, 49
stem removal from grapes, 29
Stevenage, Wine Society, 43
Stevenson, Tom, 117, 123, 126, 130, 142
still wines, 7
    champagne as, 125, 131, 132, 140
    difference from sparkling wines, 24
Stourbridge, 59–60, 68
string-rim, 50, 66, 73, 74, 75
stum/stumm, 120
sugar and yeast for secondary fermentation, 7, 23, 24, 120, 127, 129–130, 133, 134, 151, 152
sulphur, 29–30, 120
    coal's sulphurous fumes, 61
Sydney, Sir Philip, 57

tank (*Charmat; cuve close*) method, 24, 150–151
Taylor, Captain Silas, 77–78
Thames (river), 28, 65
Thévenot, Gaspard, and Thévenot bottle, 136
Thirty Years War, 138
Three Tuns (inn), 78, 79, 97
Thudichum and Dupré's *Complete manual of viticulture and Œnology*, 134–135

timber *see* wood
timelines, 15
Tower of London, 56, 82, 83, 111
transfer method and transvasage, 152
Treatise on Cider, 79
Treatise on the Preparation of Sparkling White Wines, 134
tries, 127
Trinity House, Digby as Governor of, 89, 143
Tysack family, 109

Understanding Antique Wine Bottles, 74

Van den Bossche, Willy, 72, 75
van Dyck's portraits
    Digby, 10
    Digby's wife in deathbed, 94, 95
van Helmont, Jan, 130
van Leeuwenhoek, Antonie, 130fn
Venetians, 55–56, 56, 57, 116
    Digby as plunderer of their fleet, 89
verre Anglais, 8, 22, 136, 141
Verzelini, Giacomo, 56–57, 109
vin mousseux, 129
Vinetum Angliæ, 79–80
Vinetum Brittanicum (Treatise on Cider), 79
vineyards
    Britain, mediaeval times, 27, 28, 34–35
    Champagne region, 123, 124, 126–127, 128
    European mainland in mediaeval times, 35
Vinion, John, 13, 91
Vinion, Paull, 63
Vizetelly, Henry, 141

waldglas, 22, 48, 49, 55, 61, 148
Ward, Robert, 13, 91
white wine, 42, 127, 142
    mediaeval times, 28

Wilde, Dr, 117, 119
William the Conqueror, 34
Willisel, Thomas, 121–122
Winchester House (prison), 57, 97
window glass, 53–54
wine
    bottles, 76–78
    Christie's (London auctioneers) sale of, 137–138
    fortified, 38, 120, 132
    imports, 34–39
    mediaeval winemaking, 27–31
    Merret's papers on, 117–121
    still *see* still wines
wine coopers and Merret, 10, 120–121
wine presses (grape presses), 20, 27, 28, 29, 127
Wine Society, 42–43
wine trade/merchandising, 33–43, 76–78
    with France (13th/14th C.), 35–37
    Middle Ages, 30–31, 34–37
Winter, Sir John, 14, 78, 93, 98, 109, 110–112, 144
Wollaston, 66, 67
wood (timber), 21, 50–51
    change to coal from, 59–61
    James I banning use of, 21, 51, 62, 63, 109
Wood, Thomas, 79
wound healing, Powder of Sympathy in *see* Powder of Sympathy

yeast
    Leeuwenhoek and discovery of, 130fn
    sugar and, for secondary fermentation, 7, 23, 24, 120, 127, 129–130, 133, 134, 151, 152

Zouche, Sir Edward, 62–63, 66, 75, 104

Printed in Great Britain
by Amazon